CIRC {CASTL} M6/405654
QC/178/B64
C. 1
BOWLER, M. G.
GRAVITATION AND RELATIVITY

INTERNATIONAL SERIES IN
NATURAL PHILOSOPHY

General Editor: D. TER H

D0762419

VOLUME 86

GRAVITATION
AND
RELATIVITY

Some Other Titles of Interest

ATWATER:
Introduction to General Relativity

BOWLER:
Nuclear Physics

CONSTANTINESCU & MAGYARI:
Problems in Quantum Mechanics

FURTH:
Fundamental Principles of Modern Theoretical Physics

KILMISTER:
Special Theory of Relativity

KILMISTER:
General Theory of Relativity

PATHRIA:
The Theory of Relativity, 2nd Edition

A full list of titles in the Natural Philosophy series follows index.

GRAVITATION
AND
RELATIVITY

by

M. G. BOWLER

Department of Nuclear Physics
Oxford University

PERGAMON PRESS

OXFORD · NEW YORK · TORONTO SYDNEY
PARIS · FRANKFURT

U.K.	Pergamon Press Ltd., Headington Hill Hall, Oxford OX3 0BW, England
U.S.A.	Pergamon Press Inc., Maxwell House, Fairview Park, Elmsford, New York 10523, U.S.A.
CANADA	Pergamon of Canada Ltd., P.O. Box 9600, Don Mills M3C 2T9, Ontario, Canada
AUSTRALIA	Pergamon Press (Aust.) Pty. Ltd., 19a Boundary Street, Rushcutters Bay, N.S.W. 2011, Australia
FRANCE	Pergamon Press SARL, 24 rue des Ecoles, 75240 Paris, Cedex 05, France
WEST GERMANY	Pergamon Press GmbH, 6242 Kronberg-Taunus, Pferdstrasse 1, Frankfurt-am-Main, West Germany

Copyright © 1976 Pergamon Press Ltd.

All Rights Reserved. No part of this publication may be reproduced, stored in a retrieval system, or transmitted, in any form or by any means, electronic, mechanical, photocopying, recording or otherwise, without the prior permission of Pergamon Press Ltd.

First edition 1976

Library of Congress Cataloging in Publication Data

Bowler, M G

Gravitation and relativity.

(International series in natural philosophy; v. 86)
1. Gravitation. 2. General relativity. (Physics)
I. Title.
QC178.B64 1976 530.1′1 75-42161
ISBN 0-08-020567-4
ISBN 0-08-020408-2 pbk.

In order to make this volume available as economically and rapidly as possible the author's typescript has been reproduced in its original form. This method unfortunately has its typographical limitations but it is hoped that they in no way distract the reader.

Printed in Great Britain by A. Wheaton & Co. Exeter

SELKIRK COLLEGE LIBRARY
CASTLEGAR, B. C.

CONTENTS

PREFACE

Most undergraduates in physics know a great deal of electromagnetism, special relativity and quantum mechanics by the time they have completed their course, but practically nothing about modern theories of gravitation. Yet gravity is now one of the most exciting areas of physics, both experimentally and theoretically, in addition to its unique relation with at least the large scale structure of space and time. After some forty years of relegation to the realm of cosmology, relativistic gravitation has entered astrophysics with a vengeance following the discovery of neutron stars and, very recently, compact X-ray sources. While the revival of theoretical interest in gravitation has been fed by the successful explanation of at least the qualitative aspects of stellar evolution and the apparent inevitability of gravitational collapse at the end of the life of perfectly normal stars, the technology of the second half of this century has made possible experiments of hitherto unprecedented accuracy in the field of gravitation, some qualitatively new. Einstein's theory of gravitation, general relativity, has been verified at the one per cent level, gravitational waves are being vigorously pursued and it seems quite likely that the compact X-ray source Cygnus X-1 contains a star so collapsed that even light cannot escape it, a black hole.

The interested undergraduate has no access, other than semipopular articles, to any of this excitement, lacking (as do most professional physicists) both the mathematics of general relativity and the insight and experience necessary to clothe the mathematics with physical reality. I have written this book in the hope of purveying the flavour of the physical reality but without unfamiliar mathematics. The book is based on lectures I first gave to second year undergraduates at Oxford in 1974. My audience had completed courses in electromagnetism and in special relativity, and I determined to tackle gravitation without employing on the one hand general curvilinear coordinates, or on the other the Lagrangian formulation of field theory: I have applied the elementary tools of special relativity to the problem of generalising Newton's theory of gravitation. The weakness of this approach is that it is not practicable to push beyond the weak field approximation to the Einstein field equations (although I did find a way of treating the precession of the perihelion of

Mercury) and it is not suited to a discussion of cosmology. Its strength is that it builds on material central to any physics course and should, I hope, not only deepen the understanding of these foundations but also bridge the apparent chasm separating gravitation from the rest of physics. It further has the advantage that although gravitation is treated as a classical field (and a field corresponding to mass-less spin two gravitons at that) the distortion of space-time by the presence of mass emerges as a desirable way of interpreting the results.

This approach is liable to incur odium on the grounds that it delays the inculcation of the only proper mode of thought, namely the geometric treatment of general relativity. My defence against such a charge lies mostly above, but I should remark that a field theoretical approach to gravitation leads almost inevitably to general relativity when pursued to all orders in the strength of the gravitational potential, and it should always be remembered that physics is an experimental subject and it is not impossible that a pure geometric approach might one day prove to be untenable. Anyone intending to develop a professional interest in gravitation must learn the mathematical language of general relativity: it is my hope merely to have provided a way in to the subject for those not yet so equipped, whether undergraduates or professional physicists whose training and modes of thought are subject to the same limitations as my own.

The plan of this book is as follows. In Chapter 1 special relativity is briefly reviewed, with the emphasis on the Lorentz covariance of the equations of physics. There is then a short discussion of accelerations in the framework of special relativity. Gravity first appears at the end of this chapter, when the reader is confronted with two problems. The augmenting of Newtonian gravitation by mass-energy equivalence allows a calculation of the gravitational redshift, which comes out right, and a calculation of the deflection of light by the Sun, which comes out wrong by a factor of two. A simple minded application of the principle of unique acceleration in a gravitational field (often loosely referred to as the principle of equivalence) yields precisely the same answers. The two problems are, first, how can the gravitational deflection of light be a factor of two bigger than yielded by these calculations, which get the redshift (an energy difference) right, and secondly, given that light <u>is</u> deflected by twice the result of these naive calculations, how is it possible nonetheless to maintain that there is no way of detecting a gravitational acceleration by observations within a freely falling laboratory?

Chapter 2 contains a discussion of the Eötvös-Dicke experiments which established the identity of inertial and gravitational mass, and the detailed conclusions that may be drawn from them. In Chapter 3 the equations of electrodynamics are derived by starting from the equations of electrostatics and requiring that the more general equations are Lorentz covariant: electrodynamics is used as a model for gravitation, and this machinery is applied to gravity in Chapter 4, where the field equations are set up with the conserved energy-momentum tensor as source. The observed deflection of light by the Sun is used to distinguish between various a priori possible forms for the gravitational potentials. In Chapter 5 the machinery is applied to the force laws, yielding equations of motion, and the answer to the first problem raised in Chapter 1 is provided: there is a velocity dependent force acting at right angles to the motion which deflects a particle without changing its energy. The slowing down of light in a gravitational field, together with the equations of motion in free fall, is used in Chapter 6 to find the answer to the second question raised in Chapter 1. From the point of view adopted in this book clocks are physically slowed down by a gravitational potential and measuring rods are physically contracted: effects which may be represented if desired as a distortion of space-time.

Since gravitational energy is expected to be a source of gravity, the full gravitational field equations must be nonlinear and the work of Chapters 4-6 is conducted only in the weak field approximation. Gravitational redshift, deflection of light and radar echo delay may all be discussed in these terms, but the fourth famous test of general relativity, the precession of the perihelion of Mercury, may not be. This is the subject of Chapter 7, in which the weak field equations of motion are augmented by a nonlinear term. This term is derived by requiring that the characteristic times and distances of gravitationally bound systems are affected by an external gravitational potential in the same way as atomic periods and sizes, a form of the principle of strong equivalence. Since celestial mechanics tends to be neglected in undergraduate physics courses, the phenomenon of precession and the extraction of a numerical value are treated in terms of radial oscillations superimposed on a circular orbit.

In Chapter 8 the weak field equations are again employed in discussing the nature of gravitational radiation, its generation and detection. The discovery that the radiation fields associated with the theory giving twice the Newtonian deflection of light are gauge invariant gives a theoretical motivation for this choice of theory and makes the connection with mass-less spin two gravitons.

The last two chapters are necessarily disjoint from the rest of the book. In Chapter 9 an attempt is made to establish the connection with general relativity, but the Einstein field equations are neither derived nor solved. The relation between the metric tensor and gravitational potentials is discussed, and the Schwartzschild solution in both standard and isotropic coordinates used to discuss the experimental tests of general relativity. The last chapter contains a brief discussion of black holes. The Schwarzschild solution is used to treat the propagation of light and particle motion in very strong fields and the book itself ends with a short discussion of how black holes may manifest themselves to the astronomer.

In constructing this approach to gravitation, I found the following works particularly useful: 'An alternative approach to the theory of gravitation', W.E. Thirring, Annals of Physics, 16, 96 (1961); 'Lectures on Gravitation', R.P. Feynman, (1962-63), (unpublished lecture notes); 'Gravitation without a principle of equivalence', R.H. Dicke, Rev. Mod. Phys., 29, 363 (1957). In preparing the last chapter I benefited from 'Black hole physics', R. Sexl, CERN Report TH1759 , (1973).

I am much indebted to I.J.R. Aitchison, S.J. Orebi Gann, J.C. Miller and F.N.H. Robinson for their critical reading of the draft of this book and innumerable helpful comments.

CHAPTER 1

SPECIAL RELATIVITY AND ACCELERATIONS

1.1 Special relativity in brief

The principle of relativity asserts that there is no meaningful way of defining absolute velocity. In this form it has little physical content but we can express it in more physical terms: the laws of physics are the same for all observers in uniform relative motion. Such observers inhabit inertial frames of reference: the definition of an inertial frame is that in an inertial frame a particle moves in a straight line at constant velocity unless it is acted upon by a force. This is not a circular statement: if there is a force something is present to produce it.

A coordinate transformation connects observers in different frames. If the principle of relativity is correct (and this must be decided by experiment) then a given set of equations expressing a piece of physics in one frame, subjected to such a transformation, retains its form and numerical content. In Newtonian mechanics, the equation

$$\underline{F} = m\underline{a}$$

which we may write as

$$F_i = ma_i \tag{1.1.1}$$

retains its form and numerical content under arbitrary translations and rotations of the coordinates. It also retains its form and numerical content under the Galilean transformation connecting two inertial frames

$$\begin{aligned} x' &= x - vt \qquad & z' &= z \\ y' &= y & t' &= t \end{aligned} \tag{1.1.2}$$

(augmented if so desired by translations and rotations). This transformation relates the coordinates of a given event in the primed inertial frame to the coordinates of the same event in the unprimed frame. The velocity v is the relative velocity of the two frames, for

$$\left.\frac{dx'}{dt'}\right|_{x} = -v \quad \text{and} \quad \left.\frac{dx}{dt}\right|_{x'} = +v \tag{1.1.3}$$

The spatial separation of events occurring at the same time is the same in both frames (an invariant quantity) and the acceleration of the test particle of mass m is the same regardless of which set of coordinates is used. That is for a given force any two observers in uniform relative motion measure the same acceleration. A physical law which retains its form and numerical content under a particular transformation is said to be covariant with respect to that transformation.

The equations of electromagnetism are not covariant with respect to the Galilean transformation. Consider for example the electric field of a plane wave

$$\underline{E} = \underline{E}_0 \sin(kx - \omega t) \ .$$

The phase is

$$\varphi = kx - \omega t$$

and the phase velocity is

$$\left.\frac{dx}{dt}\right|_{\varphi} = \frac{\omega}{k} = c \ . \qquad (1.1.4)$$

The magnetic field is in phase with the electric field: when the phase is zero there is no field and all observers should agree on this. The phase should be an invariant and we can use this to work out what happens to k and ω under the Galilean transformation, that is to relate the values of wavelength and frequency perceived by one observer to the values perceived by another moving relative to the first. The phase being an invariant,

$$k'x' - \omega' t' = kx - \omega t$$

must hold for all positions and times.

If x' and t' are given in terms of x and t by the Galilean transformation, we may equate coefficients of x and t and obtain

$$k' = k$$

$$vk' + \omega' = \omega$$

whence

$$\omega' = \omega(1 - \frac{v}{c})$$

and the phase velocity in the primed system is

$$c' = \frac{\omega'}{k'} = c - v \ .$$

We have obtained the Doppler shift and the expected relationship between the velocity of light in the two frames. The fact that the velocity is different itself demonstrates that the equations of electromagnetism are not covariant with respect to the Galilean transformations. The point is rubbed in by noting that as $v \to c$, $\omega' \to 0$ and in the primed frame the electric field oscillates sinusoidally with position and is constant with time. Such a field is not a solution of Maxwell's equations.

Experimentally the velocity of light, as measured with real apparatus, is a universal constant for all unaccelerated observers. We must either suppose that velocity does something funny to clocks and measuring rods or that the transformation relating the coordinates ($\underline{x}'$, t') to the coordinates ($\underline{x}$, t) is not the Galilean transformation. If the velocity of light is a universal constant regardless of the nature of the apparatus employed, these choices become operationally indistinguishable. Adopting the second choice, we can obtain the correct transformations by abstracting from Maxwell's equations one property: that the velocity of light is a universal constant.

Consider two frames of reference, each equipped with identical apparatus, but moving with velocity v relative to each other along a common x axis. (We can always make a rotation of coordinates so as to achieve this.) For convenience choose $x' = x = 0$ when $t' = t = 0$.

Let the coordinate x' be related to the unprimed coordinates through

$$x' = \alpha_{11} x + \alpha_{12} y + \alpha_{13} z + \alpha_{14} t$$

and write in general

$$x'_\mu = \alpha_{\mu\nu} x_\nu \tag{1.1.5}$$

where we employ the convention of summation over repeated indices (such a repeated index is called dummy). We expect the coefficients $\alpha_{\mu\nu}$ to depend only on the relative velocity v. This is a linear transformation connecting the coordinates of a given event (such as a wavefront reaching a specified detector) in one inertial frame with the coordinates of the same event in another inertial frame. The transformation is linear because only for a linear transformation is the unaccelerated motion of a particle in one frame seen as unaccelerated motion in the other (see Chapter 9).

The equation of an expanding wave front in the unprimed frame is

$$x^2 + y^2 + z^2 - c^2 t^2 = 0 \, . \tag{1 1.6}$$

If we choose to define coordinates $x_\mu = (x, y, z, ict)$ this can be expressed in the compact form

$$x_\mu x_\mu = 0 \, . \tag{1.1.7}$$

In the primed frame, related to the unprimed frame by

$$x'_\mu = a_{\mu\nu} x_\nu \tag{1.1.5}$$

we must have

$$x'_\rho x'_\rho = 0$$

and so

$$x'_\rho x'_\rho = a_{\rho\mu} a_{\rho\nu} x_\mu x_\nu \equiv x_\sigma x_\sigma \ .$$

This gives us at once the relation

$$a_{\rho\mu} a_{\rho\nu} = \delta_{\mu\nu} \tag{1.1.8}$$

where $\delta_{\mu\nu}$ is the Kronecker delta function, having the value 1 if $\mu = \nu$ and zero otherwise.

For transformations along the mutual x axes we can write a further relation. We have from (1.1.5)

$$x'_1 = a_{1\nu} x_\nu \ .$$

Differentiate with respect to x_4 for fixed x'_1 and get

$$a_{11} \frac{dx_1}{dx_4} + a_{14} = 0 \ . \tag{1.1.9}$$

This equation relates the transformation coefficients to the relative velocity of the two frames, and with eq. (1.1.8) determines the transformations. We first note that for motion along the mutual x axes only the x and t coordinates can be mixed without inconsistencies arising and so the relation

$$x_1'^2 + x_4'^2 \equiv x_1^2 + x_4^2$$

$$(x'^2 - c^2 t'^2 \equiv x^2 - c^2 t^2)$$

gives, on equating coefficients,

$$a_{11}^2 + a_{41}^2 = 1$$

$$a_{14}^2 + a_{44}^2 = 1$$

$$a_{11}a_{14} + a_{41}a_{44} = 0$$

which is just eq. (1.1.8) written out explicitly for this case.

If we call the velocity parameter V,

$$\frac{dx_1}{dx_4} = V = \frac{v}{ic} \ ,$$

then

$$a_{11} = a_{44} = \frac{1}{\sqrt{1 + V^2}} = \frac{1}{\sqrt{1 - \frac{v^2}{c^2}}}$$

$$a_{14} = - a_{41} = \frac{-V}{\sqrt{1 + V^2}} = \frac{-v/ic}{\sqrt{1 - \frac{v^2}{c^2}}}$$

so that

$$x_1' = \frac{x_1 - V x_4}{\sqrt{1 + V^2}}$$

$$x_4' = \frac{V x_1 + x_4}{\sqrt{1 + V^2}} \qquad (1.1.10)$$

or

$$x' = \frac{x - vt}{\sqrt{1 - v^2/c^2}}$$

$$t' = \frac{t - vx/c^2}{\sqrt{1 - v^2/c^2}} \qquad (1.1.11)$$

which are the familiar Lorentz transformations.

We now note that any quantity

$$\Delta x_\mu \, \Delta x_\mu$$

has the same numerical value in any inertial frame and is thus an invariant. The quantity Δx_μ transforms according to

$$\Delta x_\mu' = a_{\mu\nu} \, \Delta x_\nu$$

and is the prototype four-vector: its length is the prototype invariant.

Consider a particle at rest in one frame. Over any interval of time $\Delta\tau$ its spatial coordinates in that frame do not change and so

$$\Delta x_\mu' \, \Delta x_\mu' = (\Delta x_4')^2 = - c^2 \, \Delta\tau^2 \; .$$

In any other frame moving with velocity v along the x axes the particle moves a distance Δx in time Δt and so

$$\Delta x^2 - c^2 \, \Delta t^2 = - c^2 \, \Delta\tau^2$$

$\frac{\Delta x}{\Delta t} = v$ and so

$$\Delta t = \frac{\Delta\tau}{\sqrt{1 - v^2/c^2}} \qquad (1.1.12)$$

where $\Delta\tau$ is the proper time interval: the time interval elapsed in the centre of mass of the particle. This is the famous formula for time dilation. Time (as measured with real standard clocks) elapses more quickly in the laboratory frame than in the rest frame of a high energy particle passing through the laboratory.

The same result can be obtained at once from eq. (1.1.11) if we note that the x coordinate does not change in the rest frame. The recipe is quite unambiguous: the mean life of a particle moving with respect to an observer is greater than the mean life of an identical particle at rest with respect to the observer

$$\Delta t' = \frac{\Delta t - \frac{v \Delta x}{c^2}}{\sqrt{1 - \frac{v^2}{c^2}}} = \frac{\Delta t}{\sqrt{1 - \frac{v^2}{c^2}}}$$

if $\Delta x = 0$ in which case Δt is the proper time interval $\Delta\tau$.

Suppose that at time $t = t' = 0$ a high energy particle passes through a counter at $x' = x = 0$. The firing of the counter constitutes an event. Later the particle passes through a second counter and decays in it. The simultaneous firing of the second counter and the decay of the particle constitute a second event. This event also occurs at $x' = 0$ in the particle frame, but at time t'. In the laboratory frame it occurs at x, t. Then

$$0 = \frac{x - vt}{\sqrt{1 - \frac{v^2}{c^2}}} \qquad \therefore \quad x = vt$$

$$t' = \frac{t - \frac{vx}{c^2}}{\sqrt{1 - \frac{v^2}{c^2}}} = t\sqrt{1 - \frac{v^2}{c^2}} \, .$$

In time t' the point $x = 0$ has moved back a distance vt' in the particle rest frame. The distance measured in the particle rest frame between the two counters is thus

$$x' = vt\sqrt{1 - \frac{v^2}{c^2}} = x\sqrt{1 - \frac{v^2}{c^2}} \, .$$

This is the Lorentz contraction, which is however not susceptible to measurement in the direct way the time dilation is.

We now define any quantity with four components V_μ which are mixed together under the Lorentz transformations according to

$$V'_\mu = a_{\mu\nu} V_\nu \tag{1.1.13}$$

to be a four-vector. Then

$$V'_\mu V'_\mu = a_{\mu\rho} a_{\mu\sigma} V_\rho V_\sigma = V_\nu V_\nu$$

using eq. (1.1.8). The length of any four-vector is an invariant. If we have two different four-vectors the analogue of the scalar product of two ordinary vectors is

$$A'_\mu B'_\mu = a_{\mu\rho} a_{\mu\sigma} A_\rho B_\sigma = A_\nu B_\nu$$

and is also an invariant.

Thus in special relativity we can see at once that the quantities $k_\mu \left(\underline{k}, \frac{i\omega}{c}\right)$ constitute a four vector, since

$$\underline{k} \cdot \underline{x} - \omega t$$

must be an invariant. This at once gives us the relativistic Doppler effect.

We have not demonstrated that the equations of electromagnetism are Lorentz covariant (we shall do this in Chapter 3 : for a conventional treatment see ref. [1]) but we may define fields which are four-vectors $V_\mu(x_\rho)$ and four-scalars $S(x_\mu)$. We are therefore interested in the effects of differential operators on these fields. We might expect that the four quantities

$$\frac{\partial S}{\partial x_\mu}$$

make up a four-vector. We may write

$$\frac{\partial S'}{\partial x'_\mu} = \frac{\partial S}{\partial x_\nu} \frac{\partial x_\nu}{\partial x'_\mu} . \qquad (1.1.14)$$

To evaluate the quantity $\partial x_\nu / \partial x'_\mu$ we need the inverse of

$$x'_\mu = a_{\mu\nu} x_\nu \qquad (1.1.5)$$

Multiply these four equations by the appropriate $a_{\mu\rho}$ and add : this is represented by

$$a_{\mu\rho} x'_\mu = a_{\mu\rho} a_{\mu\nu} x_\nu$$

and since

$$a_{\mu\rho} a_{\mu\nu} = \delta_{\rho\nu} \qquad (1.1.8)$$

$$a_{\mu\nu} x'_\mu = x_\nu \qquad (1.1.15)$$

and

$$\frac{\partial x_\nu}{\partial x'_\mu} = a_{\mu\nu} . \qquad (1.1.16)$$

Then

$$\frac{\partial S'}{\partial x'_\mu} = a_{\mu\nu} \frac{\partial S}{\partial x_\nu} \qquad (1.1.17)$$

and the four quantities $\partial S/\partial x_\mu$ do indeed make up a four-vector : an equation of the form

$$\frac{\partial S}{\partial x_\mu} = V_\mu \qquad (1.1.18)$$

is thus Lorentz covariant.

Similarly we might expect $\partial V_\mu / \partial x_\mu$ to be invariant. Write

$$\frac{\partial V'_\mu}{\partial x'_\mu} = \frac{\partial}{\partial x'_\mu}\left\{a_{\mu\nu} V_\nu\right\} = a_{\mu\nu} \frac{\partial V_\nu}{\partial x_\rho} \frac{\partial x_\rho}{\partial x'_\mu}$$

$$= a_{\mu\nu} a_{\mu\rho} \frac{\partial V_\nu}{\partial x_\rho} = \frac{\partial V_\nu}{\partial x_\nu} \tag{1.1.19}$$

and indeed we have an invariant. An equation of form

$$\frac{\partial V_\mu}{\partial x_\mu} = S \tag{1.1.20}$$

is Lorentz covariant.

Similarly it is easy to show that

$$\frac{\partial}{\partial x_\mu} \frac{\partial}{\partial x_\mu} S$$

is a scalar field and

$$\frac{\partial}{\partial x_\mu} \frac{\partial}{\partial x_\mu} V_\nu$$

is a four-vector field. The invariant operator

$$\frac{\partial}{\partial x_\mu} \frac{\partial}{\partial x_\mu} = \nabla^2 - \frac{1}{c^2} \frac{\partial^2}{\partial t^2} \tag{1.1.21}$$

is the generalisation of the Laplacian and is called the D'Alembertian operator, frequently denoted by $\Box$.

If we accept that the equations of electromagnetism are true for all observers in inertial frames and are Lorentz covariant, then either we must discard the principle of relativity, or the equations of particle mechanics must also be Lorentz covariant. Newton's laws are covariant with respect to the Galilean transformation and must therefore be modified if the principle of relativity is to hold.

Since the proper time τ elapsed in the rest frame of a particle is an invariant, the quantities $dx_\mu/d\tau$ (where x_μ are the coordinates of that particle in any inertial frame) form a four-vector. If a particle is moving slowly, $\tau \rightarrow t$ and the first three components become the velocity. We may therefore call this quantity the four-velocity. A second differentiation provides us with a further four-vector, the four-acceleration $d^2x_\mu/d\tau^2$. Multiply the four-velocity by an invariant quantity with the dimensions of mass and we obtain a four-vector which is called the four-momentum

$$p_\mu = m_o \frac{dx_\mu}{d\tau} . \qquad (1.1.22)$$

For slow motion the first three components may be identified with the momentum in Newtonian mechanics, which is conserved. Since p_μ is a four-vector, for i particles

$$\sum_i p'_{\mu_i} = \sum_i a_{\mu\nu} p_{\nu_i} = a_{\mu\nu} \sum_i p_{\nu_i} . \qquad (1.1.23)$$

If the four-momentum is conserved for any observer in an inertial frame, it is conserved for all such observers. We may express p_μ in terms of the velocity of the particle v in any given inertial frame :

$$p_\mu = m_o \frac{dx_\mu}{dx_4} \frac{dx_4}{d\tau} \qquad (1.1.24)$$

where $x_4 = ict$ and

$$d\tau = \sqrt{1 - \frac{v^2}{c^2}}\, dt$$

thus obtaining

$$\underline{p} = \frac{m_o \underline{v}}{\sqrt{1 - \frac{v^2}{c^2}}} \qquad p_4 = \frac{m_o ic}{\sqrt{1 - \frac{v^2}{c^2}}} . \qquad (1.1.25)$$

If we define $p_4 = i\frac{E}{c}$ then

$$E = \frac{m_o c^2}{\sqrt{1 - \frac{v^2}{c^2}}} \approx m_o c^2 + \tfrac{1}{2} m_o v^2 \ldots \qquad (1.1.26)$$

in the low velocity limit. The implication of these equations is that energy has an inertial mass and conversely that inertial mass is a manifestation of energy, the two being linked by the relation $E = mc^2$. The analogue of Newton's laws is thus

$$F_\mu = m_o \frac{d^2 x_\mu}{d\tau^2} \qquad (1.1.27)$$

where F_μ is a four-force.

The invariant quantity m_o, the proper mass, is given by the square of the four-momentum

$$p_\mu p_\mu = m_o^2 \frac{dx_\mu}{d\tau} \frac{dx_\mu}{d\tau} = - m_o^2 c^2 \qquad (1.1.28)$$

or

$$p^2 - \frac{E^2}{c^2} = - m_o^2 c^2 . \qquad (1.1.29)$$

We may note that since the square of a four-vector is an invariant, only three of the components are independent. In particular, if

$$\frac{d\underline{p}}{dt} = \underline{F}$$

then

$$\frac{dE}{dt} = \underline{F} \cdot \underline{v} \quad \text{where} \quad \underline{v} = \frac{d\underline{x}}{dt} \ . \tag{1.1.30}$$

If we choose units such that $c = 1$ then $\underline{p} = E\underline{v}$ and

$$\frac{d\underline{p}}{dt} = E\frac{d\underline{v}}{dt} + \underline{v}\frac{dE}{dt} \ .$$

So

$$\underline{F} - \underline{v}\,\underline{v} \cdot \underline{F} = E\frac{d\underline{v}}{dt}$$

and so

$$\frac{d\underline{v}}{dt} = \frac{\sqrt{1 - \frac{v^2}{c^2}}}{m_o}\left\{\underline{F} - \frac{\underline{v}}{c^2}\,\underline{v} \cdot \underline{F}\right\} . \tag{1.1.31}$$

A particle exposed to a constant force in the laboratory accelerates less and less as the speed builds up.

1.2 Special relativity and accelerations

In accepting both the principle of relativity and the Lorentz transformations which were enshrined within it by Einstein, our notions of the properties of space and time have been changed. Along with these changes go changes in our notions of the properties of velocity, acceleration, force, mass, momentum and energy. Every experiment in particle physics, conducted at energies significantly greater than the rest mass energies of the particles involved, bears witness to the applicability of the Lorentz transformations to the laws of mechanics. Indeed particle physics tells us more. The tentatively constructed hypotheses concerning the interactions of the denizens of the microscopic world are all written so as to be covariant under the Lorentz transformations, and embody quantum mechanics as well as special relativity. The full range contains quantum electrodynamics (the most precisely checked of all physical theories), the theory of the weak interactions responsible for β-decay, innumerable field theories of greater or lesser physical significance, the S-matrix description of the strong interactions which hold the nucleus together and nowadays the first steps towards theories of the internal structure of the strongly interacting particles themselves. Nowhere is there evidence that the Lorentz transformations are not applicable, even when dealing with particles whose energies are hundreds of times their rest mass (protons at the Fermi National Accelerator Laboratory at Batavia, Illinois) or tens of thousands of rest masses (electrons from the linear accelerator at Stanford, California).

The Lorentz transformations link the coordinates of a given event as seen from two different inertial frames of reference. We already defined an inertial frame of reference as one in which a test particle moves with constant velocity unless acted on by a force: if we wish to be picturesque we may operationally define an inertial frame of reference as one in which it is possible to play three dimensional billiards.

A frame of reference which is being accelerated by rockets firing is clearly not an inertial frame. This has given currency to the erroneous notion that special relativity is incapable of discussing the laws of physics experienced by accelerated observers. This idea is wholly incorrect: within the postulates of special relativity we have an unambiguous recipe for discussing such observers. The crucial point is that space and time intervals as measured by different observers depend only on relative velocity and not on acceleration. This is built in to the definitions of four-velocity and four-acceleration on which covariant equations of motion involving accelerations are constructed. Thus while acceleration may break a given clock, the rate at which time elapses on a moving particle differs from the rate at which laboratory time elapses by a factor depending only on the velocity and not on the acceleration. The applicability of Lorentz covariant equations of motion to physics already reveals this.

The misconception that special relativity is helpless in the face of accelerations arises most poignantly in the so-called twin paradox. Castor is an astronaut and visits a suitably distant star, say Sirius, and returns. His journey is made at very high constant velocity, apart from brief periods of acceleration, with respect to his brother Pollux who remains at Starbase somewhere near Pluto. We may infer that on arriving home Castor finds his chronometer registers some 20 years less elapsed time than the identical one in Space Control and that his brother has aged some 20 years more than he. An alternative scenario allows Castor to accelerate at a reasonable rate, say 1 g , for half his journey, turn his spacecraft round and decelerate for the remaining half, the return trip being accomplished in like manner. The result is essentially the same.

These results are derived in the following way. Pollux uses the special relativistic formula for time dilation, which depends only on the relative velocity, to work out how much slower Castor's proper time is elapsing. He is in an inertial frame and knows that if special relativity is correct

Castor's clocks are keeping the same time with respect to an instantaneously comoving frame as his would keep if subjected to Castor's acceleration, as seen from the comoving frame. He can do experiments with such clocks and finds that for accelerations tolerable to humans a properly constructed clock measures time independently of its acceleration (although of course not independently of the velocity). Pollux thus arrives at an unambiguous answer to the problem of relative ageing.

Castor however cannot directly apply the formulae of special relativity because he has been accelerated over some parts of his journey - and he knows it because in addition to clocks he is equipped with accelerometers. These accelerations remove the symmetry between the two observers that would otherwise preclude a differential ageing, but they do not affect the proper rate of the accelerated clocks. Castor can apply the formulae of special relativity provided that he takes account of the fact that during the periods of acceleration he was continuously changing his own instantaneous inertial frame.

The twins could have worked out together the recipe for doing this before Castor ever left Starbase. The construction of such a rule book [2] is of course dependent on space-time transformations of relativity (instantaneously) not depending on accelerations : from the point of view of an observer in an inertial frame acceleration may be interpreted as the accelerated system changing inertial frames and it may be tracked by a continuously changing Lorentz transformation. This interpretation is implicit in the description of the physics of systems involving accelerations in terms of Lorentz covariant equations of motion : a description which is successful both at very high velocity and at enormous acceleration. The relativistic definitions of energy and momentum (1.1.22) hold at enormous acceleration : the implication is that the relativistic definitions of space-time intervals also hold at enormous acceleration.

It is instructive to work out the accelerations involved in quite ordinary physical systems which are well understood in terms of special relativity.

(1) <u>Atomic and nuclear structure</u>

(a) In a hydrogen atom the rate of change of velocity $\underline{v}$ of an electron is

$$\frac{v^2}{r} \sim \frac{e^2}{m_e r^2} \sim \frac{m_e e^6}{\hbar^4} . \tag{1.2.1}$$

On putting in numbers this acceleration is found to be $\sim 10^{25}$ cm s^{-2}. (The velocity of the electron is $\sim 2 \times 10^8$ cm s^{-1} and the orbital period is $\sim 10^{-16}$ s.)

The precise tests of quantum electrodynamics carried out for the hydrogen atom yield agreement between theory and measurement at a level of around 5 parts in 10^5 for the Lamb Shift and 5 parts in 10^6 for hyperfine splitting [3].

(b) Under the heading of electrons in atoms, we may include a further effect: the Thomas precession of electron spin. An electron moving through the electrostatic field of an atomic nucleus experiences in its instantaneous rest frame a magnetic field which interacts with the magnetic moment of the electron and causes a precession of the spin. The precession frequency calculated in the comoving frame is twice that observed in the laboratory. The reason is the existence of a term due entirely to special relativity and not due to any particular interaction, the Thomas precession. In the semi-classical calculation of the electron spin precession, the torque and the precession rate are evaluated in the instantaneous rest frame of the electron. The instantaneous rest frame is however precessing with respect to the laboratory frame in which the nucleus is at rest, by an amount that can be calculated straightforwardly from special relativity [4]. The evaluation of the precession rate in the instantaneously comoving inertial frame of the electron, plus the transformation back to the laboratory using the Lorentz transformaions gives the right answer for the net precession observed in the laboratory, and moreover, as it must, agrees with the answer obtained from the Dirac equation [5] in which the interaction of a spin $\frac{1}{2}$ particle with an electromagnetic field is written in an explicitly Lorentz covariant way. The acceleration as observed in the laboratory is again $\sim 10^{25}$ cm s^{-2}.

(c) Nucleons in a nucleus.

Nucleons in a nucleus are confined with $\sim 10^{-13}$ cm by the strong interactions. Their momentum is given by the uncertainty principle

$$p \sim \frac{\hbar}{R} \tag{1.2.2}$$

so their velocity is $\sim 10^{10}$ cm s^{-1}. The acceleration they experience is thus $\sim 10^{33}$ cm s^{-2}. Bound systems provide very large accelerations over large time scales.

(2) <u>Collision phenomena</u>

(a) In elastic scattering through the strong interactions a particle moving with velocity $\sim c$ can experience a change of velocity of magnitude $\sim c$ due to forces with a range of less than 10^{-13} cm. The acceleration experienced is thus $\gtrsim 10^{34}$ cm s^{-2}.

Tests of quantum electrodynamics with colliding beams of electrons of energy several GeV show no breakdown at momentum transfers corresponding to distances $\sim 10^{-15}$ cm, and hence accelerations (for 90° scattering) $\sim 10^{36}$ cm s^{-2} [6].

(b) In the bremsstrahlung process an electron radiates a photon as the result of acceleration in the field of an atomic nucleus. The change in velocity in such a process is $\sim m_e c^3/E$ for an electron energy E, corresponding to a change of momentum $m_e c$. This takes a time $\Delta t \sim \frac{r}{c}$ where r is the distance of closest approach and

$$\Delta p \sim \frac{Z e^2}{r^2} \Delta t$$

$$\sim \frac{Z e^2}{c^2 \Delta t} .$$

So that

$$\Delta t \sim \frac{Z e^2}{m_e c^3} \tag{1.2.3}$$

which for $Z = 1$ is $\sim 10^{-23}$ secs and for $Z = 100$ is $\sim 10^{-21}$ secs. For electrons of energy 1 GeV, $\Delta v \gtrsim 10^7$ cm s^{-1} and the mean acceleration is (only) $\sim 10^{30}$ cm s^{-2}.

Collision phenomena provide very high accelerations but only for very short periods.

(3) <u>Particle accelerators</u>

The accelerations we encounter here are less impressive, but are of interest since if our Lorentz covariant laws of physics broke down at high accelerations our accelerators would not work. The fact that they do work however is no precise verification of Lorentz covariant physics because accelerators are tuned to work and breakdown of Lorentz covariant physics below the 1% level would probably be tuned out. A few examples will suffice:

(a) Protons accelerated in the proton synchrotrons at CERN or Brookhaven reach an energy ~ 30 GeV (~ 30 times their rest mass energy) in ~ 1 s. Since the end velocity is $\sim c$ the mean acceleration is $\sim 3 \times 10^{10}$ cm s^{-2}, and since the particles actually experience the accelerating electric fields for $\sim 10^{-2}$ of the acceleration cycle, the peak accelerations are $\sim 10^{12}$ cm s^{-2}. The protons move in circular orbits constrained by magnetic fields and of radius $\sim 10^4$ cm so at full energy the acceleration normal to the motion is $\sim 10^{17}$ cm s^{-2}.

(b) Electrons accelerated in the Stanford linear accelerator surf ride the electric fields of waves travelling in a waveguide. The waveguide is loaded in such a way that the velocity of the travelling electric field always matches the velocity of the accelerated electron, as calculated from the equations of special relativity (1.1.30). The accelerator works: electrons emerging after two miles in the laboratory have an energy of 4×10^4 times the rest mass energy and a velocity equal to c within one part in 10^9. The velocity has been explicitly checked at the level of a few parts in 10^7, [7].

In view of these successes of special relativity in describing physics at enormous accelerations we may confidently predict that an accelerated clock runs slow with respect to an unaccelerated clock by an amount given by equation (1.1.12). This effect has been directly measured with the CERN muon storage ring.

The muon is a particle of mass 105.7 MeV/c^2 (206.8 electron masses) which appears to behave in all respects like a heavy electron. In particular the electromagnetic interactions of the muon follow quantum electrodynamics down to the smallest distances so far probed experimentally ($\sim 10^{-15}$ cm), and at this level it is still behaving like the point charge of QED. Because the muon is more massive than the electron, it may decay through the weak interactions into an electron and two neutrinos and it has a proper lifetime $\tau = 2.2 \times 10^{-6}$ s. The rate of decay of a point-like particle would appear to provide an ideal clock: within our knowledge of the weak interactions there are no internal workings to be affected by acceleration and a point-like particle can be assigned (instantaneously) a single comoving inertial frame. (This is not the case for an extended system: see Chapter 9.)

In the CERN g-2 experiment muons were injected into a ring of 5 m diameter and constrained to approximately circular orbits by a magnetic field of just over 17 kg. The object of the experiment was a precision measurement of the magnetic moment of the muon but the lifetime of the orbiting muons was a highly interesting byproduct [8]. The momentum of the stored muons was 1.27 GeV/c, their energy $\simeq$12 rest masses. If the rate at which time elapses in the accelerated muon rest frame is indeed independent of the acceleration, the lifetime observed in the laboratory should be

$$\tau\left(1 - \frac{v^2}{c^2}\right)^{-\frac{1}{2}} \simeq 12\,\tau ,$$

26 μsec instead of 2.2 μsec.

The muons were tracked over some 150 μsec, more than 10^5 revolutions. The lifetime measured in the laboratory was found to be 26.37 ± 0.05 sec, to be compared with the calculated value of 26.69 μsec. The discrepancy agrees with the estimated effects of muon losses in the course of the orbiting. The laboratory acceleration of the muons was $\sim 4 \times 10^{18}$ cm s^{-1}. This work thus provided the experimental coup de grace to the interminable twin paradox (which is dead but won't lie down).

Lorentz covariant laws of physics work beautifully not only at velocities which are within one part in 10^9 of that of light, but also at accelerations in excess of 10^{30} cm s^{-2}. Mass, energy, momentum and velocity are all (instantaneously) acceleration independent. Electric and magnetic fields act on an accelerating particle according to prescription and we have direct experimental evidence that the rate at which proper time elapses is acceleration independent.

Thus if we want to work out the physics of an accelerated laboratory, as seen by an observer in that laboratory, all we have to do is to work out the physics in a given inertial frame and then transform to the instantaneously comoving frame to find what our accelerated observer will instantaneously see. This of course is exactly what is done to discuss the physics of accelerated laboratories within the framework of Newtonian physics and leads to the introduction of centrifugal and coriolis forces. We must however use Lorentz covariant physical laws and the Lorentz transformations, at least if we are studying any phenomena involving high velocities. We must also recognise that in addition to the continuous transition between comoving frames that occurs in acceleration, for extended systems different parts may be in different comoving frames. It is the fitting together of all these different comoving frames as a function of space and time coordinates that constitutes the generalisation of special relativity into general relativity. But general relativity also embodies a theory of gravity. We have not mentioned gravity at all yet, because we have not so far attempted to write down Lorentz covariant laws of gravitation.

1.3 Acceleration and gravity

Physical theories embedded in special relativity correctly describe physical systems even at enormous accelerations, provided one is careful to view these systems from an inertial frame of reference, in which accelerometers read zero and three dimensional billiards can be played. A physicist in an accelerated laboratory knows he is being accelerated, because his accelerometers do not read zero and he cannot play three dimensional billiards. He cannot use the Lorentz transformations without supplementing them with rules for taking account of his continuously changing inertial frame.

Suppose however this physicist is accelerating in a gravitational field, under free fall. His accelerometers read zero and he can play three dimensional billiards. Gravity has a singular property — or at least a property that seems singular to a physicist mostly brought up on the physics of electromagnetism and structures held together by electromagnetic forces. Everything is accelerated in a gravitational field at the same rate, and consequently in a small laboratory in free fall there is no internal way of detecting the acceleration. This is now a matter of common experience, at least seen through the dark glass of the television screen. Conversely, a physicist in a laboratory with an accelerometer (for example, a mass on a spring) reading 1 g has no internal way of telling whether his laboratory is at rest on the surface of the earth or being blasted (by silent and vibration free engines) beyond the orbit of Jupiter. Gravity thus has a singular link with acceleration and inertia — or perhaps we should put it the other way around and say that acceleration and inertia are intimately linked with gravitation.

These statements are based on Newton's laws, appropriate to low velocity phenomena, and the Eötvös-Dicke experiments which have failed to detect any differential gravitational acceleration of different objects, at an accuracy $\sim$ one part in 10^{11} - 10^{12} (see Chapter 2). (The experiences of astronauts, though stimulating, do not constitute any very precise test of this statement.) Since we are concerned with gravitation and relativity, we should investigate whether this principle of equivalence will hold for high velocities too. The highest velocity available is c, so we examine the effects of both gravity and accelerations, of the kind produced by rockets, on light.

We will consider first an accelerating rocket, in which the acceleration a is not too different from g, just to keep things simple. There are two problems we will work out to first order in the acceleration: the frequency shift of light due to acceleration and the departure of light from rectilinear propagation.

Light is emitted from a source in the nose of the rocket at time $t = 0$, as measured in the frame comoving with the nose at this instant. It heads towards the tail of the rocket with constant velocity c in this same frame. Because the rocket is accelerating, in time t it picks up a velocity with respect to this original frame of at, and travels a distance $\frac{1}{2}at^2$ — these Newtonian expressions are quite adequate for the low changes of velocity involved. If the distance between the source and receiver is h, the time taken to traverse this distance at velocity c is $t = \frac{h}{c}$ and hence the velocity of the receiver with respect to the frame in which the emitter was instantaneously at rest is approximately $\frac{ah}{c}$ at the instant the light is received (see Fig. 1.3.1). The frequency of the light in the comoving receiver frame is thus Doppler shifted on reception at the tail of the rocket by an amount

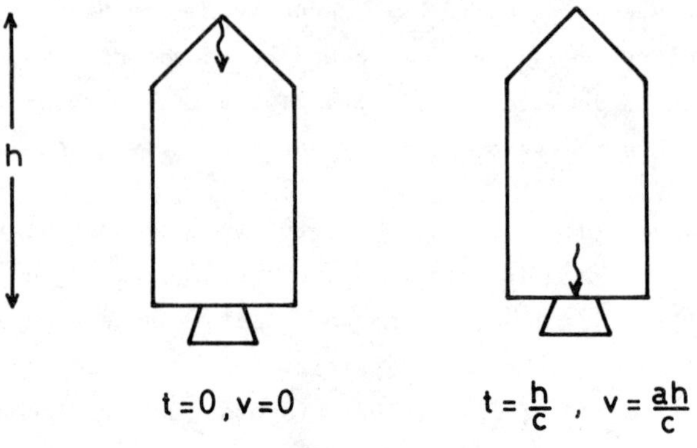

Fig. 1.3.1

$$\frac{\Delta\nu}{\nu} = \frac{ah}{c^2} . \tag{1.3.1}$$

Corrections due to the path travelled in the emission frame being slightly less than h, Lorentz contraction and time dilation are all second order in small quantities. If the acceleration a is g and h is 10 m, then

$$\frac{\Delta\nu}{\nu} \simeq 10^{-15} . \tag{1.3.2}$$

In a quantum picture the same result of course obtains because the transformations for energy and frequency are the same: both energy and frequency transform as the fourth component of a four-vector.

Next consider a beam of light emitted at right angles to the acceleration in the appropriate instantaneous comoving frame. In this frame it travels in a straight line with velocity c, covering a distance $\Delta x = c\Delta t$ in time Δt. In this time however the rocket has advanced a distance

$$\Delta y = \tfrac{1}{2}a(\Delta t)^2 = \tfrac{1}{2}a\left(\frac{\Delta x}{c}\right)^2 . \tag{1.3.3}$$

If the light is to pass through a set of holes in the rocket structure at various Δx, the holes must not lie on a straight line but regress towards the tail of the rocket along the parabola (see Fig. 1.3.2)

$$\Delta y = \tfrac{1}{2}a\left(\frac{\Delta x}{c}\right)^2 .$$

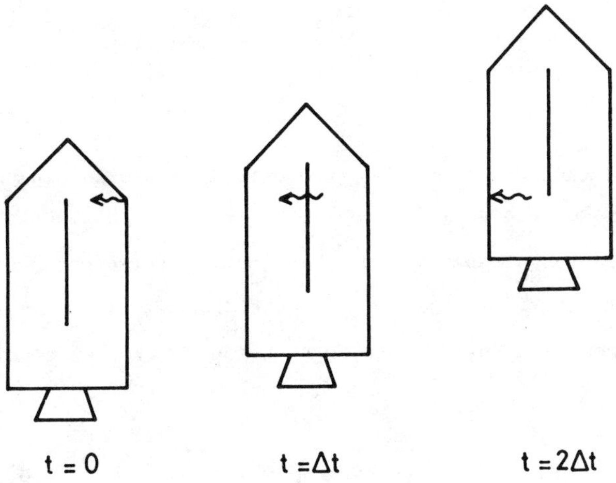

Fig.1.3.2 The figure illustrates the parabolic trajectory of a photon in the accelerating frame of reference, corresponding to a linear trajectory in a specified inertial frame

The angle the light has been deflected through (as seen by the physicist in in the space vehicle) is then

$$\alpha = \frac{\partial \Delta y}{\partial \Delta x} = \frac{a \Delta x}{c^2} \qquad (1.3.4)$$

We can apparently say that if the principle of equivalence applies to light, of velocity c, then in a gravitational field g electromagnetic waves are shifted in frequency in falling a distance h by

$$\frac{\Delta \nu}{\nu} = \frac{gh}{c^2} \qquad (1.3.5)$$

and light is deflected through an angle

$$\Delta\alpha = \frac{g}{c^2} \Delta x \qquad (1.3.6)$$

in travelling a distance Δx at right angles to the gravitational acceleration.

We may now see how this ties up with an alternative approach in which we directly consider light in a gravitational field. We need the principle of equivalence in a slightly different form this time. In special relativity the inertial mass of a system (a particle, atom, light pulse or whatever you like) is equal to its total energy content divided by c^2. A light pulse of energy E thus has inertial mass $\frac{E}{c^2}$ and if we are right in thinking that it is indeed always the inertial mass that governs the gravitational interaction, then the gravitational energy of a light pulse in a gravitational potential φ will be $\frac{E}{c^2}\varphi$, and it will experience an acceleration $-\underline{\nabla}\varphi$.

The momentum picked up sideways in travelling a distance Δx at right angles to the field is thus

$$\Delta p = \frac{E}{c^2} \underline{\nabla}\varphi \frac{\Delta x}{c} \qquad (1.3.7)$$

and since the momentum is $\frac{E}{c}$ the angle of deflection is

$$\Delta\alpha = \underline{\nabla}\varphi \frac{\Delta \mathbf{x}}{c^2} = g \frac{\Delta \mathbf{x}}{c^2} \qquad (1.3.8)$$

which agrees with the result we obtained by watching a rocket with proper acceleration g from an inertial frame.

We can also get an answer for the gravitational frequency shift. An atom at rest has internal energy E and in a gravitational potential φ this becomes $E\left(1 + \frac{\varphi}{c^2}\right)$. We have conservation of energy and so for a transition between two levels, separated by ΔE in the absence of a potential, the photon energy is $\Delta E\left(1 + \frac{\varphi}{c^2}\right)$. The difference in energy between a photon emitted at a potential φ_1 and a photon emitted at φ_2 is

$$\Delta E_2 - \Delta E_1 = \Delta E\left(\frac{\varphi_2 - \varphi_1}{c^2}\right) = \Delta E \frac{gh}{c^2} \qquad (1.3.9)$$

which agrees with the Doppler shift arguments for an accelerating rocket. It is important to note that the argument we have just constructed depends on conservation of energy and the weight of binding energy $\mathcal{E}$ being equal to $g\frac{\mathcal{E}}{c^2}$: that is, binding energy behaves inertially and gravitationally like (negative) mass. We shall examine the best evidence for this (the Eötvös-Dicke experiments) in Chapter 2.

1.4 Measurements of the gravitational frequency shift

The gravitational deflection of light has not been observed in the laboratory, but the gravitational frequency shift has been verified to the 1% level using the Mössbauer effect [9]. 14.4 KeV photons from ^{57}Fe ($\tau = 10^{-7}$ sec) were employed, going both up and down a path of 22.5 m at Harvard. The fractional difference in frequency between photons going up (red shift) and photons coming down (blue shift) is thus 4.905×10^{-15}, detected by achieving resonance via the Doppler effect: the source must move at a velocity of 7×10^{-5} cm s^{-1} to re-establish resonance. The difficulty of the experiment is brought out by noting that the fractional line width, for recoilless emission, is $\sim 10^{-12}$. The measurement was in fact made by investigating the difference in asymmetry of the resonance line as a function of velocity for red and blue shifted photons. The result obtained was 0.9990 ± 0.0076 of the expected effect, the error being purely statistical, with an additional possible uncertainty of ~ 0.01, being the linear sum of all contributing systematic errors. The conclusion is that the gravitational frequency shift (1.3.9) is verified at the 1% level [10].

The gravitational redshift suffered by a photon in climbing up a 22.5 m tower is only 2.5×10^{-15}. Larger redshifts are available astronomically. A photon escaping from the sun will have its frequency redshifted by an amount $\frac{GM_\odot}{R_\odot c^2} = 2 \times 10^{-6}$, while a photon escaping from a white dwarf of mass $\sim M_\odot$ and radius $\sim 10^4$ km would be redshifted by $\sim 1.5 \times 10^{-4}$. Observations of the solar gravitational redshift are in accord with expectation and have reached a precision $\sim 5\%$ [11]. A comparison of measured and expected redshift for white dwarfs requires in addition to measurement knowledge of the mass and radius of the star: there is agreement but the precision is only $\sim 15\%$ [12]. The terrestrial Mössbauer experiment is thus by far the most precise measurement of gravitational frequency shift, the astronomical measurements checking more crudely the first order prediction for stronger fields.

1.5 The gravitational deflection of light

While the gravitational deflection of light has not been measured in the laboratory, the deflection of light by the Sun was first measured in 1919 and constitutes one of the great tests of Einstein's theory of gravity, general relativity. We have already worked out the angle light is deflected in going

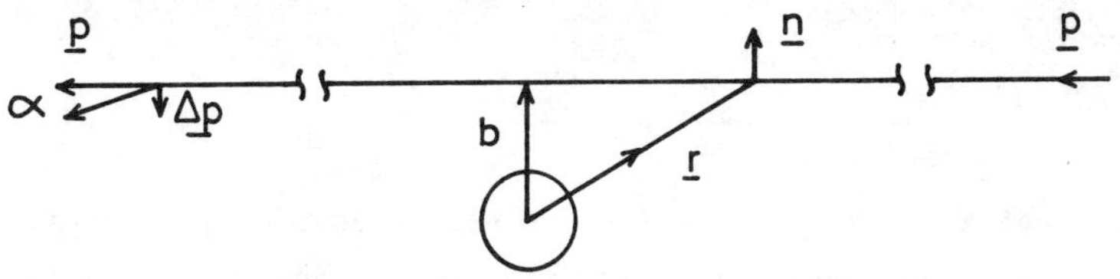

Fig. 1.5.1 The transverse momentum acquired by a photon in passing the Sun is calculated by integrating the component of gravitational force at right angles to the approximately linear trajectory

a small distance Δx through a locally constant gravitational field: let us use this result to calculate the deflection of light by the Sun. Because the deflection is very small we do not need to calculate the orbit but just calculate the change of momentum by integrating the normal component of the force along a straight line (Fig. 1.5.1).

$$\frac{dp_n}{dt} = -\frac{E}{c^2}\frac{GM_\odot}{r^2}\frac{\underline{r}\cdot\underline{n}}{r} \qquad (1.5.1)$$

so

$$\Delta p_n = \frac{E}{c^2}\int \frac{GM_\odot}{r^2}\frac{\underline{r}\cdot\underline{n}}{r}\,dt \qquad (1.5.2)$$

$$dt = \frac{dx}{c} \quad \text{and} \quad p = \frac{E}{c}$$

so

$$\alpha(b) = \frac{\Delta p_n}{p} = \frac{GM_\odot}{c^2}\int \frac{\underline{r}\cdot\underline{n}}{r^3}\,dx \; . \tag{1.5.3}$$

The integral can be evaluated directly or by using Gauss' theorem, for it represents $\frac{1}{2\pi b}$ multiplied by the normal flux out of an infinitely long cylinder of radius b. Thus

$$\alpha(b) = \frac{2GM_\odot}{bc^2} \tag{1.5.4}$$

and

$$\alpha(R_\odot) = \frac{2GM_\odot}{R_\odot c^2}$$

for light grazing the limb of the Sun.

In this derivation we have used the relativistic relation between energy and momentum for a photon (or light pulse), set the force equal to $-\frac{E}{c^2}\underline{\nabla}\varphi$, using the relativistic relation between energy and inertial mass, and equated the force to the rate of change of momentum. Precisely the same answer is obtained by assuming that light is not observed to be deflected as it passes across a box freely falling in a gravitational field and constructing the path by fitting together such freely falling boxes, for such boxes are observed to be accelerating at a rate

$$\frac{d^2y}{dt^2} = -\frac{GM_\odot}{r^2}\,\frac{\underline{r}\cdot\underline{n}}{r} \tag{1.5.5}$$

as seen from outside the solar system and hence, as seen from outside the solar system the light beam has an equation of motion

$$\frac{d^2y}{dx^2} = -\frac{GM_\odot}{c^2r^2}\,\frac{\underline{r}\cdot\underline{n}}{r} \; . \tag{1.5.6}$$

The angle of deflection is

$$\frac{dy}{dx} = \frac{GM_\odot}{c^2}\int \frac{\underline{r}\cdot\underline{n}}{r^3}\,dx \tag{1.5.7}$$

so $\alpha(b) = 2GM_\odot/bc^2$ once more.

Putting in numbers, $M_\odot = 1.99\times10^{33}$ gm, $R_\odot = 6.96\times10^{10}$ cm, $G = 6.67\times10^{-8}$ cgs units

$$\alpha(R_\odot) = 4.245\times10^{-6}\ \text{rad}$$

$$= 0.875'' \; .$$

Experimental observations are all consistent with a value

$$\alpha(R_\odot) = 1.75''$$

the value which almost everyone knows is predicted by Einstein's theory of gravity through the relation

$$\alpha(R_\odot)_{GR} = \frac{4GM_\odot}{R_\odot c^2} \tag{1.5.8}$$

which is just twice the Newtonian value we have calculated.

Experimentally $\alpha(R_\odot)$ has been measured in two ways :

(1) By photographing the star field around the Sun during a total eclipse and measuring the displacement of star images relative to those on plates taken when the same stars appear in the night sky (at a remove of ~6 months). The results have a spread on $\alpha(R_\odot)$ of ~ 1.3" - 2.7" and may be taken as consistent with $\alpha(R_\odot)_{GR}$ to within an error ~25% [13]. The problems are: (i) observations are limited to $\alpha(>2R_\odot)$ because of glare from the corona, (ii) total eclipses do not usually enshadow observatories with big telescopes: the diffraction image size for a 10 cm instrument is $\sim 5 \times 10^{-6}$ rad, (iii) it is necessary to compare separate plates taken and developed independently, at an interval of some months.

(2) Long base line interferometry has been used to determine the change in the apparent position in the sky of the quasi-stellar radio source 3 C 279 during its annual occultation by the Sun. The relative phase of the signals received by two radiotelescopes is monitored and continually compared with the relative phase of the signals received from the quasar 3 C 273 which is not occulted and is 9.5° away from 3 C 279 . The wavelengths used lie in the range 3-15 cm and most baselines in the range 1-20 km . The relative phase change for a baseline d is $\sim \frac{2\pi d}{\lambda}\alpha$ so a shift in phase of one radian corresponds to $\alpha \sim \frac{\lambda}{2\pi d}$ which for 10 cm radiation and a 20 km baseline is $\sim 10^{-6}$ radians. Most measurements have been restricted to values of the impact parameter $\geqslant 3R_\odot$ because of refraction due to free electrons in the solar corona. Because the refractive index of the solar corona is frequency dependent and the expected deflection of radio waves by gravity is not, simultaneous work at two or more different frequencies allows smaller impact parameters to be used, while the use of baselines of thousands of kilometres should eventually permit the determination of $\alpha(R_\odot)$ to very much better than 1%. Early results were consistent with $\alpha(R_\odot) = 1.75''$ with errors ≈10% [13].

The most recent analysis of the occultation of 3 C 279 has yielded $\alpha = (0.99 \pm 0.03)\,\alpha_{GR}$, obtained with an interferometer baseline of 845 km [14].

Dual frequency interferometry has very recently been applied to obtain the most accurate result to date [15]. Instead of the usual quasar 3 C 279, three almost co-linear radio sources were used, 0119 + 11, 0116 + 08 and 0111 + 02. The occulted source was 0116 + 08 and the outer two, lying respectively $\sim 4^{\circ}$ and 6° away on opposite sides of the apparent path of the sun, were observed to eliminate local transient effects. The baseline was 35 km and the deflection found to be

$$\alpha = (1.015 \pm 0.011)\ \alpha_{GR} .$$

The observations of the deflection of electromagnetic radiation in the gravitational field of the Sun are thus beautifully in accord with the predictions of Einstein's theory, made ~ 60 years ago, and totally inconsistent with the value calculated by augmenting Newtonian gravity with mass-energy equivalence. In most of our subsequent work we shall assume that the deflection of light is twice the Newtonian value.

1.6 An apparent paradox

We have arrived at a situation in which our simple calculations are not only at variance with observation but lead to apparently paradoxical conclusions. We have calculated both the gravitational redshift and the deflection of light by the Sun in two different ways: by using Newtonian gravity augmented by mass-energy equivalence and by using the equivalence of effect of a gravitational field and an acceleration. Both methods agree, for each phenomenon, but while we get an answer in accord with experiment for the first effect, redshift, we are wrong by a factor of two for the gravitational deflection of light. The observed value for the deflection of light suggests that while no gravitational redshift is observed in a box in free fall, light will be curved by an amount depending on the local gravitational field, violating the principle of equivalence. Everyone has been told however that general relativity is founded on the principle of equivalence and yields both

$$\frac{\Delta\nu}{\nu} = \frac{gh}{c^2} \quad \text{and} \quad \alpha(R_\odot) = \frac{4GM_\odot}{R_\odot c^2} ,$$

statements we have found to be apparently incompatible.

We can see a possible way out of the paradox presented by our using Newtonian gravitation to compute successfully the energy change of a photon in a gravitational field and unsuccessfully to compute the deflection in the solar gravitational field. The component of force we integrated to give the deflection of the photon was at right angles to the motion. If gravity provides an additional force always at right angles to the motion which is equal to the Newtonian force for $v = c$, and negligible for $v \ll c$, then the result is explicable because a force at right angles to the motion does no work and hence cannot change the energy. The motion of a charged particle in a magnetic field provides an example, and magnetism is a relativistic effect of electric fields. We shall therefore search for some sort of gravitational analogue of magnetism and leave the apparent violation of the principle of equivalence alone, for the time being. We shall embark on this search in Chapter 4, after examining the evidence for the equivalence of inertial and gravitational mass in Chapter 2 and studying the relation between electromagnetism and relativity in Chapter 3.

References

[1] See for example J.D. Jackson, Classical Electrodynamics, (Wiley 1962) Ch. 11 ; H. Muirhead, The Special Theory of Relativity, (Macmillan 1973) Ch. 6.

[2] See for example E.F. Taylor and J.A. Wheeler, Spacetime Physics, (Freeman 1963) Exercises 27, 49, 81.

[3] See S.J. Brodsky and S.D. Drell, Ann. Rev. Nucl. Sci., 20, 147 (1970) B.E. Lautrup et al., Phys. Rep., 3C, 196 (1972).

[4] See for example E.F. Taylor and J.A. Wheeler, Spacetime Physics (Freeman 1963), Exercise 103; J.D. Jackson, Classical Electrodynamics (Wiley 1962), Ch. 11.5.

[5] See for example, L.I. Schiff, Quantum Mechanics, (3rd ed. McGraw-Hill 1968), sect. 52; L. Mandl, Quantum Mechanics (2nd ed., Butterworth 1957), Ch. X.

[6] J-E. Augustin et al., Phys. Rev. Lett., 34, 233 (1975).

[7] Z.G.T. Guiragossián et al., Phys. Rev. Lett., 34, 335 (1975).

[8] F. Combley and E. Picasso, Phys. Rep., 14C, 3 (1974) J. Bailey et al., Phys. Lett., 55B, 420 (1975).

[9] See The Mössbauer Effect (ed. H. Frauenfelder) Benjamin 1962.

[10] R.V. Pound and J.L. Snider, Phys. Rev., 140, B788 (1965).

[11] J. Brault (1962). A brief discussion may be found in C.W. Misner, K.S. Thorne and J.A. Wheeler, Gravitation, (Freeman 1973) P1059. J.E. Blamont and F. Roddier, Phys. Rev. Lett., 7, 437 (1961).

[12] J.L. Greenstein and V. Trimble, Ap. J., 149, 283 (1967).

[13] See C.W. Misner, K.S. Thorne and J.A. Wheeler, Gravitation (Freeman 1973) Ch. 40. S. Weinberg, Gravitation and Cosmology (Wiley 1972) Ch. 8.5.

[14] C.C. Counselman et al., Phys. Rev. Lett., 33, 1621 (1974).

[15] E. Fomalont and R. Sramek, reported in Physics Today, 28, No 4, 17 (1975) E. Fomalont and R. Sramek, Ap. J., 199, 749 (1975).

SELKIRK COLLEGE LIBRARY
LEGAR, B. C.

CHAPTER 2
THE EOTVOS-DICKE EXPERIMENTS

2.1 Gravitational and inertial mass

We have so far been assuming that the source of the gravitational interaction is the inertial mass of a system: before we proceed further we must examine the evidence for this, for it is possible that the gravitational mass M_g and the inertial mass M_i are not the same. Thus we should write the Newtonian law of gravity as

$$F = G \frac{M_{g1} M_{g2}}{r^2} \tag{2.1.1}$$

and equate such a force to $M_i a$ where a is the resulting acceleration. The equation of motion of a simple pendulum would then be written

$$M_i \frac{d^2\theta}{dt^2} = - M_g \frac{g}{\ell} \theta \tag{2.1.2}$$

and its angular frequency

$$\omega = \sqrt{\frac{M_g}{M_i} \frac{g}{\ell}} .$$

If $M_g = \gamma M_i$, with γ a universal constant, then γ is set equal to 1 in the definition of G. If we define G by the relation

$$F = G \frac{M_{i1} M_{i2}}{r^2} \tag{2.1.3}$$

then if γ varied with position in space, with time, or from material to material, this would appear operationally as a variation of G, assumed in Newtonian theory to be a universal constant. If γ is different for different materials, then the period of a pendulum will depend on the material of the bob,

$$\omega = \sqrt{\gamma \frac{g}{\ell}} .$$

2.2 The Eötvös-Dicke experiments

Experiments to test the universality of γ therefore search for a difference in the acceleration of ~~two~~ different objects falling in the same gravitational field. The greatest precision has been achieved in the null experiments first carried out by Eötvös at the beginning of the century and in the early 1960's by Dicke and his school. The Dicke group looked for a differential acceleration between two different objects falling in the gravitational field of the Sun. The principles are easily understood. Consider an ideal earth with its rotational axis at 90° to the orbital plane. On the equator, two masses of different materials but equal weight, are suspended at opposite ends of the arm of a torsion balance, pointing North-South. At dawn each mass experiences a force

$$M_g \frac{GM_{g\odot}}{r^2}$$

to the East. To keep it in orbit a force $M_i v^2/r$ is required, directed to the East, where v is the orbital velocity of the earth. If the quantity γ is different for the two masses on the torsion arm, one will fall towards the Sun a little faster than the other, unless restrained by a twist in the suspension. The effective forces acting on each mass, as seen in the rest frame of the apparatus, are thus a gravitational force

$$F_g = M_g \, G \frac{M_{g\odot}}{r^2}$$

to the East and a centrifugal force

$$F_i = \frac{M_i v^2}{r}$$

acting to the West. At dusk, 12 hours later, the gravitational forces act towards the West and the centrifugal forces towards the East. If the ratio of M_g and M_i were not the same for both masses, there would result an oscillating torque about the suspension with a period of 24 hours (Fig.2.2.1).

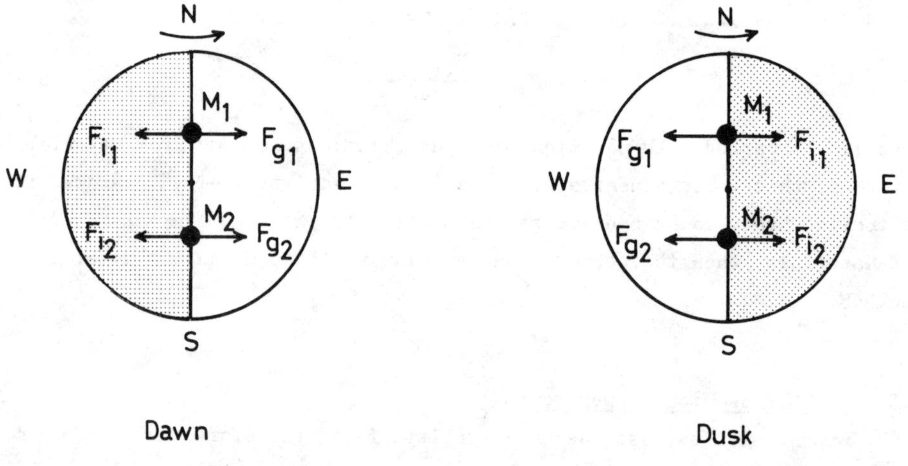

Fig.2.2.1 An illustration of the principles of the Dicke experiment to test the equivalence of inertial and gravitational mass

This oscillating torque would produce an angular oscillation about the suspension with a period of 24 hours. In the most sensitive of Dicke's experiments the two masses were of gold and aluminium and such an oscillation was absent at a level ($\sim 10^{-9}$ rad) implying equality of the constant of proportionality γ for gold and aluminium at a level of one part in 10^{11}: equality to within 3 parts in 10^{11} at the 95% confidence level is quoted. Eötvös searched for an imbalance between gravitational forces due to the earth's gravitational field and the centrifugal forces caused by the earth's rotation. The apparatus had to be rotated periodically (whereas Dicke let the rotation of the earth do it for him) and deflection of the torsion balance was looked for with optical lever, telescope and the naked eye, instead of optical lever, telescope and electronics as in Dicke's experiments. Nonetheless, Eötvös established the equality of γ for a wide variety of substances to a few parts in 10^9. Reference [1] contains a detailed account of Dicke's experiments and discussion of earlier work establishing the equality of inertial and gravitational mass, especially that of Eötvös. A measurement similar in principle to those of Dicke and his school has been carried out by Braginsky and Panov [2], who report equality of γ for aluminium and platinum within 0.9 in 10^{12} (95% confidence limit).

2.3 Implications of the null result of Eötvös-Dicke experiments

We now examine the consequences which flow from this null result. Remember that in calculating the gravitational redshift using gravity (rather than the equivalence of gravity and acceleration) we assumed that the energy of a system (such as an atom) in a gravitational field depended only on its total energy content (that is, its mass) and the potential and that conservation of energy held locally in the transitions. The Braginsky version of the Eötvös-Dicke experiment yields the result

$$|\gamma_{Pt} - \gamma_{A\ell}| < 10^{-12} .$$

We know the inertial mass of an atom to be composed of the following terms (at least)

(1) Electron rest mass
(2) Electron kinetic energy
(3) Electron potential energy (electrostatic energy)
(4) Proton and neutron rest masses
(5) Proton and neutron kinetic energy
(6) Proton and neutron potential energy due to nuclear forces
(7) The electrostatic energy of the nucleus.

Now

$$\gamma = \frac{M_g}{M_i} = \frac{\sum_j M_g^j}{\sum_j M_i^j} = \frac{\sum_j \gamma_j M_i^j}{\sum_j M_i^j} = \frac{\sum_j \gamma_j M_i^j}{M_i} \qquad (2.3.1)$$

where M_g^j is the gravitational mass of a particular component and M_i^j is the inertial mass of a particular component. Then

$$\delta\gamma_j = \frac{M_i}{M_i^j}\,\delta\gamma . \qquad (2.3.2)$$

Define $\eta = \gamma_{Pt} - \gamma_{A\ell}$ so that the anomaly due to a single term, $\delta\gamma_j$ is given by

$$|\delta\gamma_j| = \left| \frac{\delta\eta}{\dfrac{M^j_{i\,Pt}}{M_{i\,Pt}} - \dfrac{M^j_{i\,A\ell}}{M_{i\,A\ell}}} \right| = \left| \frac{10^{-12}}{\dfrac{M^j_{i\,Pt}}{M_{i\,Pt}} - \dfrac{M^j_{i\,A\ell}}{M_{i\,A\ell}}} \right| \qquad (2.3.3)$$

unless of course there are grossly implausible cancellations.

The biggest difference between Pt and Aℓ is in the neutron/proton ratio. The compositions of Pt and Aℓ atoms are

Platinum:	78 protons, 78 electrons, 117 neutrons
Aluminium:	13 protons, 13 electrons, 14 neutrons.

The fractional weight of neutrons is thus 0.65 for Pt and 0.52 for Aℓ.

The anomaly in γ due to neutrons is thus less than one part in 10^{11} and similarly for protons. If there is an anomaly associated with the neutron-proton mass difference, which is about 0.15%, then it must be less than one part in 10^9. Similarly, if electrons have a ratio of gravitational to inertial mass different from unity, the difference must be less than one part in 10^8.

We may now turn our attention to the contributions of binding and kinetic energy rather than the rest mass energy of atoms. On putting together electrons, protons and neutrons to make an atomic nucleus and its retinue of electrons, these constituents pick up kinetic energy but lose a greater amount of potential energy in the formation of a stable system, which has less mass than the rest masses of the constituents. The mean binding energy is the modulus of the sum of these two terms, divided by the mass number (number of nucleons in the nucleus) of the atom. The mean binding energies of Pt and Aℓ differ by about 0.3 MeV — approximately 3×10^{-4} of the nucleon rest mass energy. The anomaly in γ due to binding energy is thus $\leqslant 3 \times 10^{-9}$: the gravitational mass-equivalent of binding energy is equal to the inertial mass-equivalent of binding energy to within 3 parts in 10^9. We can now break this down into its component parts.

Nuclear electrostatic self-energy

The electrostatic self-energy $\mathcal{E}$ of a nucleus is quite well represented by the expression

$$\mathcal{E} = \frac{3}{5}\frac{Z^2e^2}{R} \qquad R = R_0 A^{\frac{1}{3}} , \tag{2.3.4}$$

A being the atomic mass number, Z the atomic number and $R_0 \simeq 10^{-13}$ cm. The fraction of inertial mass contributed is thus (+) $\sim 8 \times 10^{-3}$ for Pt and (+) 3×10^{-3} for $A\ell$, so $\delta\gamma_{electrostatic} \leqslant 2 \times 10^{-10}$.

Nuclear binding

The binding energy of a nucleus is given quite accurately by the semi-empirical formula [3]

$$B(Z,A) = \alpha A - \beta A^{\frac{2}{3}} - \gamma\left(\frac{N-Z}{A}\right)^2 - \mathcal{E} \pm \delta \tag{2.3.5}$$

where $\mathcal{E}$ is the electrostatic term we have already evaluated, and δ is an oscillating correction for nuclei of even A. The fraction of mass represented by the first term is the same for Pt and $A\ell$: it is a first approximation to the difference between kinetic and potential energy. The second term is a correction due to nucleons near the surface not interacting with the same number of neighbours as those deep within the nucleus. We may take the second term to provide a measure of the quantity $\delta\gamma$ due to the strong interaction potential. The constant β is about 18 MeV, so the fraction of inertial mass contributed by the term in β is $\sim 3 \times 10^{-3}$ for Pt and $\sim 6 \times 10^{-3}$ for $A\ell$, and for the strong interactions $\delta\gamma \leqslant 3 \times 10^{-10}$. The limit on $\delta\gamma$ due to nucleon kinetic energy is probably about the same: we can make an estimate of it from the third term which essentially represents excess kinetic energy forced on nucleons by the Pauli principle, in a nucleus where the number of protons and neutrons is different. The fraction of inertial mass contributed by this term is $\sim 10^{-2}$ in Pt and $\sim 10^{-5}$ in $A\ell$: we thus find a limit on $\delta\gamma$ from nucleon kinetic energy $\sim 10^{-10}$. (The constant γ in the semi-empirical mass formula, not to be confused with the ratio of gravitational and inertial mass, is about 23 MeV.)

Atomic structure

The binding energy of the atomic electrons may be estimated from the Thomas-Fermi model of the atom [4]

$$B_e(Z) = 15.73\ Z^{7/3}$$

and is ~ 6 KeV for $A\ell$, 0.41 MeV for Pt. Electron binding energy thus contributes a fractional inertial mass $\sim 2 \times 10^{-7}$ in $A\ell$, $\sim 2 \times 10^{-6}$ in Pt. Then

$\delta\gamma_{Be} \leqslant 5 \times 10^{-7}$ with the same sort of limit on the potential and kinetic energies of the electrons separately.

Gravitational self-energy

The Eötvös-Dicke-Braginsky experiments do not provide a measure of the ratio of the gravitational mass-equivalent and inertial mass-equivalent of gravitational self-energy. The gravitational self-energy of a uniform sphere of mass M, radius R is $-\frac{3}{5}\frac{GM^2}{R}$ which for Pt is $\sim 10^{-38}$ ergs. The fractional contribution to the inertial mass is thus $\sim 3 \times 10^{-38}$.

Antimatter

This is more exotic, but of interest because of the frequently encountered speculation that matter and antimatter might be mutally repulsive gravitationally. The inertial mass of antimatter (positrons, antiprotons ...) is certainly positive so such a speculation implies negative gravitational mass. Now atoms contain temporarily virtual electron-positron pairs, which manifest themselves through corrections to the electrostatic potential [5], and if the positron were repelled by the gravitational field of ordinary matter, the gravitational mass would be anomalous by an amount [6]

$$\sim m_e \left(\frac{Ze^2}{\hbar c}\right)^2$$

where the dimensionless quantity $(Ze^2/\hbar c)^2$ is the order of magnitude of the probability of finding a pair in the field of the nucleus. We can see that we expect a dependence on Ze^2 (rather than Z^2e^2) by a simple argument similar to that often used to calculate the range of nuclear forces due to the π-meson (pion) — but caveat emptor.

If we separate a pair of electrons in the Coulomb field of the nucleus an energy $Ze^2\Delta\left(\frac{1}{r}\right)$ is available. If this is equal to $2m_ec^2$ we may expect virtual pairs to be important. The time scale Δt over which energy is uncertain to ΔE is given by $\Delta E\,\Delta t \sim \hbar$ and $\Delta r < c\Delta t \simeq \frac{\hbar}{m_ec}$. This is also the Compton wavelength of the electron — if we cannot localise the electrons better than this, we may set $r \sim \Delta r \sim \frac{\hbar}{m_ec}$ and expect pairs to be important if

$$\frac{Ze^2}{\hbar c} \sim 1 .$$

This sort of effect is known as vacuum polarisation — the effect of the virtual electrons is real in electromagnetism because their transient presence alters the effective charge distribution near the source of an electric field [4] and contributes to the Lamb shift and the anomalous magnetic moment of the electron. We may therefore have confidence that the gravitational and inertial masses of positrons are the same at a level

$$\delta\gamma_{e^+} < 10^{-12} \frac{M_{Pt}}{m_e} \left(\frac{Z_{Pt}}{137}\right)^{-2} \sim 2 \times 10^{-6} .$$

Similar effects are expected to occur for the strong interactions: virtual proton-antiproton pairs (and pairs of other strongly interacting particles) should have a transient existence. We have no well developed theory from which the size of possible effects can be computed, but the depth of the average strong interaction potential in the nucleus is about 10^{-1} $M_p c^2$ so the probability of finding nucleon-antinucleon pairs is perhaps $\sim 10^{-2}$. We should be safe in inferring from the Braginsky experiment that $\delta\gamma_{\bar{p}_{etc}} < 10^{-6}$ and still have an enormous margin for error: we can be confident that antimatter does not fall upwards. The most precise test of the gravitational equivalence of matter and antimatter is found in $K^0 - \bar{K}^0$ interference phenomena [7].

Weak interactions

We have considered potential energy from the strong, electromagnetic and gravitational interactions. In the first two cases the Eötvös-Dicke-Braginsky experiments allow us to conclude with great precision that the associated inertial and gravitational masses are identical. In the case of gravitational energy we can learn nothing from these experiments. The only other known class of interactions is the class of weak interactions responsible for nuclear beta decay. The fundamental interaction responsible for β decay may be represented diagrammatically as

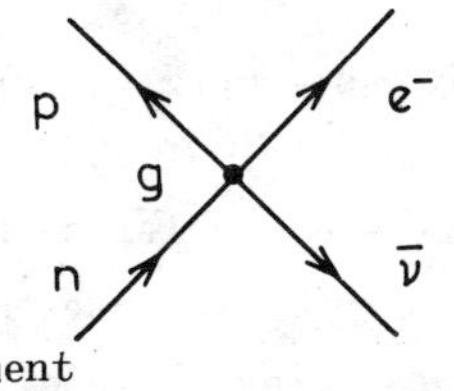

with an associated matrix element

$$g \int \varphi_e^* \varphi_{\bar{\nu}}^* \Psi_p^* \Psi_n \, dV \qquad (2.3.6)$$

where the Fermi constant $g \simeq 1.4 \times 10^{-49}$ erg cm^3 [3].

There is evidence in nuclear physics for a similar interaction between two nucleons

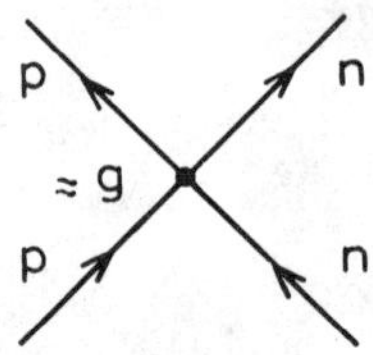

and the energy associated with this is given by the diagonal matrix element

$$g \int \Psi_p^* \Psi_p \Psi_n^* \Psi_n \, dV \sim \frac{g}{V} \qquad (2.3.7)$$

where V is the nuclear volume. The weak interaction energy of just one pair of nucleons in $A\ell$ is thus $\sim 10^{-12}$ ergs $\simeq 6 \times 10^{-7}$ MeV corresponding to $\sim 3 \times 10^{-11}$ of the inertial mass.

If the total energy goes as the number of nucleon pairs, the fraction of the inertial mass contributed by the weak interactions is $\sim 10^{-8}$, in all nuclei. If we suppose a 1% imbalance in this fraction between $A\ell$ and Pt then η would be $\sim 10^{-10}$ if the weak interaction energy had no gravitational mass. If the imbalance was due to one pair of nucleons in $A\ell$, the corresponding η would be $\sim 3 \times 10^{-11}$. Thus the Braginsky version of the Eötvös-Dicke experiment suggests $\delta\gamma_{weak} < 10^{-1} - 10^{-2}$, but there is not much margin for error.

In summary, the ratio of gravitational to inertial mass of binding energy is unity to within 3 parts in 10^9. Potential energy and kinetic energy considered separately have $\gamma = 1$ to a rather better accuracy. Little can at present be said about the gravitational properties of the weak interactions, and nothing about the gravitational properties of gravitational energy.

Conservation of energy and the mass-energy relation of special relativity are verified in nuclear physics at a level of about one part in 10^4. This has been done by comparing the masses of nuclei as determined through mass spectroscopy with corresponding energy differences measured through reaction kinematics. With the equivalence of gravitational and inertial mass of energy established at the level of a few parts in 10^9 and the inertial mass equivalent verified at one part in 10^4 [8], it would indeed have been astonishing if the gravitational redshift experiment had failed to yield the expected answer at the 1% level.

We may now confidently accept the non-relativistic relation

$$F = G \frac{M_1 M_2}{r^2}$$

where M_1 and M_2 are inertial masses and attempt to find the relativistic laws of gravitation. Our position is analogous to that of a martian physicist who has access to neither metals nor lodestone , but has nonetheless studied electrostatics and discovered the Lorentz transformations of special relativity. Before attempting to find the relativistic laws of gravitation, we will first attempt this martian task of finding the full laws of electrodynamics from Coulomb's law and special relativity.

References

[1] P.G. Roll, R. Krotkov and R.H. Dicke, Annals of Physics, 26, 442 (1964).

[2] V.B. Braginsky and V.I. Panov, JETP, 34, 464 (1971).

[3] See for example M.G. Bowler, Nuclear Physics, (Pergamon 1973).

[4] See for example L.D. Landau and E.M. Litschitz, Quantum Mechanics, (2nd ed., Pergamon 1965) sect. 70.

[5] See for example J.D. Bjorken and S.D. Drell, Relativistic Quantum Mechanics, (McGraw-Hill 1964).

[6] L.I. Schiff, Phys. Rev. Lett., 1, 254 (1958).

[7] M.L. Good, Phys. Rev., 121, 311 (1961). For a general discussion of K^0 decay phenomena, see D.H. Perkins, Introduction to High Energy Physics (Addison-Wesley 1972) sect. 4.11.

[8] A.H. Wapstra, In Handbuch der Physik XXXVIII/1 (Springer-Verlag 1958) 1 .

CHAPTER 3

MARTIAN ELECTRODYNAMICS

3.1 Fields

We will imagine a martian physicist who in the dry atmosphere of Mars has acquired an excellent knowledge of electrostatics. Without having discovered electromagnetism, he nonetheless decides to attempt to measure the velocity of Mars through the hypothetical medium which supports light waves, obtains a null result with interferometers with arms of both equal and unequal length and is driven to discover the Lorentz transformations. He finds a principle of relativity philosophically attractive and so embarks on the task of finding a set of physical laws which are covariant under the Lorentz transformations and reduce in the low velocity limit to the laws of electrostatics. For the rest of this chapter we shall follow the hypothetical reasoning of this hypothetical martian.

Abstracting from Coulomb's law we define the electric field of a point charge q to be given by

$$\underline{E}_q(\underline{r}) = q\,\frac{\underline{r}}{r^3} \tag{3.1.1}$$

The surface integral of $\underline{E}_q$ is

$$\int_S \underline{E}_q \cdot \underline{n}\; dS = q \int \frac{\underline{r}\cdot\underline{n}}{r^3}\, dS = 4\pi q \tag{3.1.2}$$

if q is inside the closed surface (and zero if it is not). If many point charges are present we use the principle of superposition of electric fields to write

$$\int_S \underline{E}\cdot\underline{n}\; dS = 4\pi \sum_{\text{inside}} q_i \tag{3.1.3}$$

and defining the local charge density as

$$\rho = \lim_{V\to 0} \frac{\sum_V q_i}{V} \tag{3.1.4}$$

we have

$$\int_S \underline{E}\cdot\underline{n}\; dS = 4\pi \int_V \rho\; dV$$

where V is the volume bounded by the closed surface S. Now

$$\int_S \underline{E} \cdot \underline{n}\, dS = \int_V \underline{\nabla} \cdot \underline{E}\, dV = 4\pi \int_V \rho\, dV$$

and since the volume considered is arbitrary we have

$$\underline{\nabla} \cdot \underline{E} = 4\pi\rho \ . \tag{3.1.5}$$

With the further definition

$$\underline{E} = -\underline{\nabla}\varphi$$

where φ is a scalar function of position, we reach Poisson's equation

$$\nabla^2\varphi = -4\pi\rho \tag{3.1.6}$$

and this is a convenient form of the laws of electrostatics from which to start the construction of electrodynamics.

This equation is clearly not covariant under the Lorentz transformations, because the operator ∇^2 contains only derivatives with respect to x , y and z . We need to replace ∇^2 by an operator with well defined properties under the Lorentz transformation, in which x , y , z and ict all enter symmetrically and which in the static limit reduces to ∇^2 . We therefore replace the Laplacian operator ∇^2 by the D'Alembertian

$$\Box = \frac{\partial^2}{\partial x^2} + \frac{\partial^2}{\partial y^2} + \frac{\partial^2}{\partial z^2} - \frac{1}{c^2}\frac{\partial^2}{\partial t^2} = \nabla^2 - \frac{1}{c^2}\frac{\partial^2}{\partial t^2} \qquad (1.1.21);(3.1.7)$$

and this will give us an equation

$$\Box\,\varphi = -4\pi\rho \ . \tag{3.1.8}$$

In a region of space where ρ is zero, we have

$$\nabla^2\varphi - \frac{1}{c^2}\frac{\partial^2\varphi}{\partial t^2} = 0 \tag{3.1.9}$$

which has solutions

$$\varphi = f(\underline{k} \cdot \underline{x} - \omega t)$$

where

$$k^2 = \frac{\omega^2}{c^2} \ .$$

The replacement of ∇^2 by the appropriate operator in which x , y , z and ict enter symmetrically has at once yielded us an equation with a free field solution which propagates as a wave at the speed of light.

Our hypothetical martian would undoubtedly be tempted to identify his theoretically discovered electric waves with light, but at once he discovers a difficulty. The electric field associated with the potential φ is

$$\underline{E} = -\underline{\nabla}\varphi = -\underline{k}\, f' \ .$$

For example, if

$$\varphi = \varphi_0 e^{i(\underline{k} \cdot \underline{x} - \omega t)}$$

then

$$\underline{E} = -\underline{k}\,\varphi_0\, e^{i(\underline{k} \cdot \underline{x} - \omega t)} \quad .$$

This electric field is in the direction of propagation of the wave and there is only one polarisation. If our martian is acquainted with optics he will know that light has two internal degrees of freedom, two independent polarisations.

We therefore examine more critically the content of the equation

$$\Box \varphi = -4\pi\rho \, .$$

Since the D'Alembertian is a four-scalar operator, the whole equation is only covariant under the Lorentz transformations if φ and ρ have the same transformation properties. If φ is a four-scalar, then ρ must be a four-scalar. The charge of a particle is

$$q = \int \rho \, dV = \int \rho \, dx \, dy \, dz \quad .$$

If such a particle is moving past the observer who uses coordinates $\underline{x}'$, t', then he sees a charge

$$q' = \int \rho' \, dx' \, dy' \, dz' \quad .$$

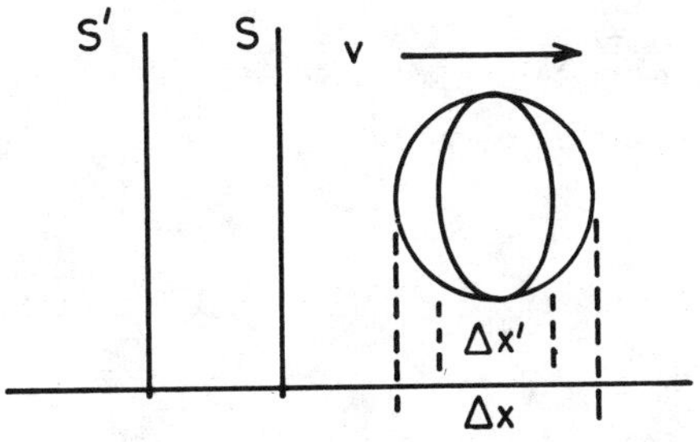

Fig. 3.1.1 Lorentz contraction of a moving charge distribution

If ρ is a four-scalar, then $\rho' = \rho$. We have $\Delta y' = \Delta y$, $\Delta z' = \Delta z$ but

$$\Delta x' = \Delta x \sqrt{1 - \frac{v^2}{c^2}} \quad ,$$

the Lorentz contraction. Thus

$$q' = q \sqrt{1 - \frac{v^2}{c^2}} \; :$$

if ρ is a four-scalar, charge is a function of velocity and tends to zero as $v \to c$. If our martian has already done experiments with high energy electrons (goodness knows how) he will reject this possibility: if he has not then he has an addition to his list of crucial experiments to be performed when technically feasible.

If we reject the identification of φ and ρ with four-scalars, we may consider the possibility that

$$\Box \varphi = -4\pi\rho$$

relates the same components of two four-vectors. If ρ is the fourth component of a four-vector it will transform like an interval of time. Since

$$\Delta t' = \Delta t \left(1 - \frac{v^2}{c^2}\right)^{-\frac{1}{2}} \quad , \quad \rho' = \rho \left(1 - \frac{v^2}{c^2}\right)^{-\frac{1}{2}}$$

and the charge density increases so as to compensate for the Lorentz contraction of a charge distribution and leave the charge invariant,

$$q' = \int \rho' \, dx' \, dy' \, dz' = \int \left(\frac{\rho}{\sqrt{1 - \frac{v^2}{c^2}}} \right) \left(dx \sqrt{1 - \frac{v^2}{c^2}} \right) dy \, dz = q \, . \quad (3.1.10)$$

This is an attractive possibility and becomes more attractive if we allow our martian knowledge of conservation of charge. Conservation of charge plus the general statement that the laws of physics are the same for observers in all inertial frames requires local conservation of charge: we may not have charge disappearing at one point in space and simultaneously appearing at another point, because different observers do not agree on the simultaneity or otherwise of events at different coordinates in space. Thus charge only disappears from a volume of space by flowing in a continuous way through the bounding surface. Local conservation of charge is contained in the equation of continuity

$$\underline{\nabla} \cdot \underline{J} + \frac{\partial \rho}{\partial t} = 0 \qquad (3.1.11)$$

where $\underline{J}$ is current density, $\underline{J} = \rho \underline{v}$.

Now ict is the fourth component of a four-vector. If we identify $(\underline{J}, ic\rho)$ as the components of a four-vector, the equation of continuity can be written in the manifestly covariant form

$$\frac{\partial \mathcal{J}_\mu}{\partial x_\mu} = 0 \qquad (3.1.12)$$

where

$$\mathcal{J}_\mu = (\underline{J}, ic\rho) \quad , \quad x_\mu = (\underline{x}, ict)$$

and $\partial \mathcal{J}_\mu / \partial x_\mu$ is shorthand for

$$\frac{\partial \mathcal{J}_1}{\partial x_1} + \frac{\partial \mathcal{J}_2}{\partial x_2} + \frac{\partial \mathcal{J}_3}{\partial x_3} + \frac{\partial \mathcal{J}_4}{\partial x_4}$$

in the convention of summing repeated (dummy) indices. The equation of continuity thus takes the form of a statement that the four-divergence of the four-current is zero and we have a relation valid for all observers in inertial frames. With the charge density the fourth component of a four-vector, we have charge an invariant and locally conserved.

For this vector theory of electricity and related phenomena we write

$$\Box A_\mu = -\frac{4\pi}{c}\mathcal{J}_\mu \qquad A_\mu = (\underline{A}, i\varphi) \tag{3.1.13}$$

and the equation for φ is recovered on setting $\mu = 4$. There are now four internal degrees of freedom in the field, but only two polarisations for light. These four degrees of freedom are reduced to three on noting that since $A_\mu \sim \mathcal{J}_\mu$ and

$$\frac{\partial \mathcal{J}_\mu}{\partial x_\mu} = 0$$

we must also have

$$\frac{\partial A_\mu}{\partial x_\mu} = 0 \tag{3.1.14}$$

thus providing one constraint. We still have one more degree of freedom than obtains for light however, but the gauge invariance of the equations decouples one of these degrees of freedom: we shall leave the discussion of this until Chapter 8.

3.2 Forces

The next task confronting our martian is an investigation of the forces acting on charged particles: so far he has only obtained the potentials. In his experimentally accessible electrostatics he has for the energy density of a charge distribution in an external potential φ the three-scalar quantity

$$\mathcal{E} = \rho\varphi \tag{3.2.1}$$

and a force density given by

$$\underline{\mathfrak{F}} = -\rho\underline{\nabla}\varphi . \tag{3.2.2}$$

He therefore expects the forces in the general case to involve the four-scalar function

$$\mathcal{I} = -\frac{1}{c}\mathcal{J}_\mu A_\mu , \tag{3.2.3}$$

corresponding to a current-current interaction between two charge distributions

$$\mathcal{I} \sim \frac{1}{c^2}\mathcal{J}_\mu(x)\,\mathcal{J}'_\mu(x')\,f(x-x') , \tag{3.2.4}$$

and the derivative operator $\partial/\partial x_\nu$. Since the force on a pointlike particle in the static case is

$$\underline{F} = \frac{d\underline{p}}{dt} = -q\underline{\nabla}\varphi \tag{3.2.5}$$

in the general case we must have

$$\frac{d\underline{p}}{dt} = \frac{1}{c}J_\mu\underline{\nabla}A_\mu \tag{3.2.6}$$

plus perhaps other terms, where

$$J_\mu = \int \mathcal{J}_\mu\, d^3x \quad ; \quad \underline{J} = q\underline{v} , \quad J_4 = icq .$$

This equation is certainly not written in a covariant way: we have three-vectors on both sides and differentiation with respect to t. We may start to write it in a covariant way by replacing the $\underline{p}$ and $\underline{\nabla}$ three-vectors with four-vectors:

$$\frac{dp_\nu}{dt} = \frac{1}{c} J_\mu \frac{\partial}{\partial x_\nu} A_\mu + \text{other terms}\ . \tag{3.2.7}$$

Now $p_\nu = \left(\underline{p}\, ,\, i\frac{E}{c}\right)$ so the generalised force equation yields the rate of change of energy as well as the rate of change of momentum:

$$\frac{dE}{dt} = \underline{F}\cdot\underline{v} \quad \text{when} \quad \frac{d\underline{p}}{dt} = \underline{F} \qquad (1.1.30) \tag{3.2.8}$$

so that when $\underline{v} = 0$, $\frac{dE}{dt} = 0$.

Since $\underline{J} = (q\underline{v}\, ,\, icq)$, only J_4 is non-zero when $\underline{v} = 0$. So we have

$$\left.\frac{dp_4}{dt}\right|_{\underline{v}=0} = \frac{J_4}{c}\frac{\partial A_4}{\partial x_4} + \text{other terms} = 0\ . \tag{3.2.9}$$

This is achieved by setting

$$\frac{dp_\nu}{dt} = \frac{1}{c} J_\mu \left\{\frac{\partial A_\mu}{\partial x_\nu} - \frac{\partial A_\nu}{\partial x_\mu}\right\} \tag{3.2.10}$$

which is identically zero for $\mu = \nu$.

We still have the problem of a differentiation with respect to t alone on the left hand side, and we may be worried because $J_\mu = \int \mathcal{J}_\mu d^3\mathbf{x}$: it is $\mathcal{J}_\mu$ that is the four-vector. We can however rewrite this equation in a manifestly covariant way:

$$\frac{dp_\nu}{dt} = \frac{dp_\nu}{d\tau}\frac{d\tau}{dt}$$

$$J_\mu = q\frac{dx_\mu}{dt} = q\frac{dx_\mu}{d\tau}\frac{d\tau}{dt}$$

so that

$$\frac{dp_\nu}{d\tau} = \frac{q}{c}\frac{dx_\mu}{d\tau}\left\{\frac{\partial A_\mu}{\partial x_\nu} - \frac{\partial A_\nu}{\partial x_\mu}\right\}\ . \tag{3.2.11}$$

This equation is clearly covariant, for the charge of the particle q is an invariant, the four-momentum of the particle, p_ν, is a four-vector and so is its derivative with respect to the scalar quantity τ, the proper time measured in the instantaneous rest frame of the particle. The derivative of the four-coordinate x_μ of the particle, $dx_\mu/d\tau$, is also a four-vector, the four-velocity, and so the right-hand side is a four-vector formed by taking two four-vectors and a four-vector operator. The quantity

$$\frac{\partial A_\mu}{\partial x_\nu} - \frac{\partial A_\nu}{\partial x_\mu}$$

is frequently denoted by $F_{\nu\mu}$ and is called the electromagnetic field tensor [1].

We can obtain the force laws in a familiar form by working out the space-like and time-like components of equation (3.2.10), obtaining

$$\begin{aligned} \frac{d\underline{p}}{dt} &= q\left\{-\underline{\nabla}\varphi - \frac{1}{c}\frac{\partial \underline{A}}{\partial t} + \frac{1}{c}\underline{v} \times \underline{\nabla} \times \underline{A}\right\} \\ \frac{dE}{dt} &= q\underline{v} \cdot \left\{-\underline{\nabla}\varphi - \frac{1}{c}\frac{\partial \underline{A}}{\partial t}\right\} \quad . \end{aligned} \tag{3.2.12}$$

The term in $\underline{\nabla} \times \underline{A}$ provides a force always at right angles to the velocity and so does not affect the energy of a charged particle: this is the magnetic term. With the definitions

$$\left.\begin{aligned} \underline{B} &= \underline{\nabla} \times \underline{A} \\ \underline{E} &= -\left\{\underline{\nabla}\varphi + \frac{1}{c}\frac{\partial \underline{A}}{\partial t}\right\} \end{aligned}\right\} \tag{3.2.13}$$

we have the familiar expressions

$$\left.\begin{aligned} \frac{d\underline{p}}{dt} &= q\left\{\underline{E} + \frac{1}{c}\underline{v} \times \underline{B}\right\} \\ \frac{dE}{dt} &= q\underline{v} \cdot \underline{E} \end{aligned}\right\} \tag{3.2.14}$$

We may calculate the acceleration of a particle by using the relation between momentum and energy

$$\underline{p} = m\underline{v} = \underline{v}\frac{E}{c^2}$$

which gives

$$m\frac{d\underline{v}}{dt} = \frac{d\underline{p}}{dt} - \frac{\underline{v}}{c^2}\frac{dE}{dt} \tag{3.2.15}$$

so that

$$\frac{d\underline{v}}{dt} = \frac{q}{m}\left\{\underline{E} - \underline{v}\frac{\underline{v} \cdot \underline{E}}{c^2} + \frac{1}{c}\underline{v} \times \underline{B}\right\} \tag{3.2.16}$$

where

$$m = \frac{E}{c^2} = \frac{m_0}{\sqrt{1 - \frac{v^2}{c^2}}} \quad .$$

Thus by searching for a set of relativistically covariant laws that reduce in the low velocity limit to the familiar laws of electrostatics, our hypothetical martian has uncovered the laws of electrodynamics. We will parallel this reasoning in Chapters 4 and 5 in an attempt to find the laws of gravitation of which Newtonian gravitation represents only the low velocity limit.

3.3 The Lagrangian formalism: for experts

The four-scalar interaction $\frac{1}{c} \mathcal{J}_\mu A_\mu$ has the dimensions of energy density but does not transform as an energy density. It can however be interpreted as an invariant Lagrangian density corresponding to the interaction of the current density $\mathcal{J}_\mu$ with the external field A_μ. Integration over the space variables provides, for a point-like particle, the interaction Lagrangian

$$\frac{q}{c} \frac{dx_\mu}{d\tau} A_\mu . \qquad (3.3.1)$$

The equations (3.2.11) may be obtained at once from the invariant Lagrangian

$$L = \frac{1}{2} m_o \frac{dx_\mu}{d\tau} \frac{dx_\mu}{d\tau} + \frac{q}{c} \frac{dx_\mu}{d\tau} A_\mu \qquad (3.3.2)$$

inserted in the covariant version of the Euler-Lagrange equations [2]

$$\frac{d}{d\tau} \frac{\partial L}{\partial u_\mu} - \frac{\partial L}{\partial x_\mu} = 0 \qquad (3.3.3)$$

where

$$u_\mu = \frac{dx_\mu}{d\tau} .$$

References

[1] J.D. Jackson, Classical Electrodynamics, (Wiley 1962) Chapter 11.

[2] For example, H. Goldstein, Classical Mechanics, (Addison-Wesley 1950) Section 6.6.
See also L.D. Landau and E.M. Lifshitz, The Classical Theory of Fields, (2nd ed., Pergamon 1962) Section 23.

CHAPTER 4

RELATIVISTIC GRAVITATIONAL FIELDS

4.1 The gravitational Poisson equation

We first obtain the gravitational version of Poisson's equation. Abstracting from Newton's law of gravity, we define the gravitational force field due to a point mass M as

$$\underline{H}(\underline{r}) = - GM \frac{\underline{r}}{r^3} . \tag{4.1.1}$$

The surface integral of this force field $\underline{H}$ is

$$\int \underline{H} \cdot \underline{n} \, dS = - 4\pi GM . \tag{4.1.2}$$

To obtain the equivalent of Poisson's equation, we take a volume ΔV containing a total mass M composed of a number of smaller masses M_i each generating a field

$$\underline{H}_i(\underline{r}) = - GM_i \frac{\underline{r} - \underline{r}_i}{|\underline{r} - \underline{r}_i|^3} \tag{4.1.3}$$

$$\int \underline{H}_i \cdot \underline{n} \, dS = - 4\pi G M_i . \tag{4.1.4}$$

If $\underline{H} = \Sigma \underline{H}_i$ then

$$\int \underline{H} \cdot \underline{n} \, dS = - 4\pi G \, \Sigma M_i \tag{4.1.5}$$

but it is important to note the assumption that gravitational fields obey the principle of superposition, that is, they add vectorially. This is certainly justified in a weak field approximation, but we must beware when considering strong fields.

We define the mass density ρ as

$$\rho = \lim_{\Delta V \to 0} \frac{\Sigma M_i}{\Delta V} ,$$

when using Gauss's theorem we have

$$\int \underline{H} \cdot \underline{n} \, dS = \int \underline{\nabla} \cdot \underline{H} \, dV = - 4\pi G \int \rho \, dV \tag{4.1.6}$$

or

$$\underline{\nabla} \cdot \underline{H} = - 4\pi G \rho . \tag{4.1.7}$$

We now define a gravitational scalar potential h so that $\underline{H} = - \underline{\nabla} h$ whence

$$\nabla^2 h = 4\pi G \rho \tag{4.1.8}$$

which is the gravitational equivalent of Poisson's equation.

In order to construct a covariant version of this equation we first make the replacement

$$\nabla^2 \to \Box$$

$$\Box h = 4\pi G\rho . \qquad (4.1.9)$$

In empty space we have

$$\left(\nabla^2 - \frac{1}{c^2}\frac{\partial^2}{\partial t^2}\right) h = 0 \qquad (4.1.10)$$

which has solutions which are waves travelling with velocity c : we are at once led to anticipate the existence of gravitational waves propagating with velocity c .

4.2 The properties of the source of gravitational fields

Since $\Box$ is a scalar operator, the gravitational potential h must have the same transformation properties as the mass density ρ : to find out more about the gravitational potentials we must study the properties of ρ. We happily have more experimental information than our hypothetical martian inventing electrodynamics : we know that particles of velocity c (light) are deflected in a gravitational field.

Suppose that the gravitational potential h is a four-scalar. Then ρ would be a four-scalar. However, if this were the case we would have

$$M' = \int \rho' \, dx' \, dy' \, dz' = \int \rho \sqrt{1 - \frac{v^2}{c^2}} \, dx \, dy \, dz = \sqrt{1 - \frac{v^2}{c^2}} \, M \qquad (4.2.1)$$

and the gravitational mass would tend to zero as $v \to c$ and a particle of velocity c would not be deflected by a gravitational field.

Could h be the fourth component of a four-vector h_μ, as the scalar electromagnetic potential is the fourth component of A_μ ? Such an identification requires ρ to be the fourth component of a four-vector and hence M to be an invariant. The inertial mass of a particle is not an invariant. The proper mass is an invariant, but for a photon the proper mass is zero and so again light would not be deflected by a gravitational field. Indeed a particle of arbitrary proper mass m_0 would not be deflected as its velocity $\to c$, since the gravitational force would be $F \propto m_0$ while the momentum

$$p = mv = \frac{m_0 v}{\sqrt{1 - \frac{v^2}{c^2}}}$$

so that

$$\frac{\Delta p}{p} = \int \frac{F \, dt}{p} \to 0$$

as $v \to c$: an electron is deflected through ever smaller angles in the Coulomb field of a nucleus as $v \to c$.

We must consider the properties of the mass density ρ more carefully, and turn once more for help to the conservation laws. We have the equivalence of inertial mass and energy and we have conservation of energy. Relativistic covariance implies local conservation of energy which we express through an equation of continuity:

$$\underline{\nabla} \cdot \underline{\mathcal{J}}_{\mathcal{E}} + \frac{\partial \mathcal{E}}{\partial t} = 0 \tag{4.2.2}$$

where $\underline{\mathcal{J}}_{\mathcal{E}} = \underline{v}\mathcal{E}$ and $\mathcal{E}$ is energy density. We must beware of making the identification of the quantities $(\mathcal{J}_{\mathcal{E}}, ic\mathcal{E})$ with a four-vector, because we know that energy and momentum make up a four-vector. Each component of momentum is also conserved, and if $\underline{\mathcal{P}}$ denotes the momentum density, then for the x component of momentum we have

$$\underline{\nabla} \cdot (\underline{v}\,\mathcal{P}_x) + \frac{\partial}{\partial t}\mathcal{P}_x = 0 \tag{4.2.3}$$

and for the ith component of momentum

$$\underline{\nabla} \cdot (\underline{v}\,\mathcal{P}_i) + \frac{\partial \mathcal{P}_i}{\partial t} = 0$$

which we may write as

$$\frac{\partial}{\partial x_j} v_j\, \mathcal{P}_i + \frac{\partial \mathcal{P}_i}{\partial t} = 0 \tag{4.2.4}$$

We can combine these equations with the equation for energy conservation and write (in the absence of external forces)

$$\frac{\partial \mathcal{T}_{\mu\nu}}{\partial x_\mu} = 0 \tag{4.2.5}$$

where $\mathcal{T}_{\mu\nu}$ is the energy-momentum density tensor.

Since $\underline{\mathcal{P}} = \underline{v}\,\frac{\mathcal{E}}{c^2}$ $(= \underline{v}\,\rho)$ we can write out $\mathcal{T}_{\mu\nu}$ explicitly.

$$\mathcal{T}_{\mu\nu} = \mathcal{E} \begin{pmatrix} \frac{v_x v_x}{c^2} & \frac{v_y v_x}{c^2} & \frac{v_z v_x}{c^2} & \frac{i\,v_x}{c} \\ \frac{v_x v_y}{c^2} & \frac{v_y v_y}{c^2} & \frac{v_z v_y}{c^2} & \frac{i\,v_y}{c} \\ \frac{v_x v_z}{c^2} & \frac{v_y v_z}{c^2} & \frac{v_z v_z}{c^2} & \frac{i\,v_z}{c} \\ \frac{i\,v_x}{c} & \frac{i\,v_y}{c} & \frac{i\,v_z}{c} & -1 \end{pmatrix} \tag{4.2.6}$$

or setting $(\underline{v}, ic) = \tilde{v}_\mu$

$$\mathcal{T}_{\mu\nu} = \frac{\tilde{v}_\mu \tilde{v}_\nu}{c^2} \mathcal{E} . \qquad (4.2.7)$$

This is only a shorthand however, $\tilde{v}_\mu$ as defined here is not a four-vector, because $\tilde{v}_\mu = dx_\mu/dt$: the corresponding four-vector is $u_\mu = dx_\mu/d\tau$.

A tensor $t_{\mu\nu}$ has the transformation

$$t'_{\mu\nu} = a_{\mu\rho} a_{\nu\sigma} t_{\rho\sigma}$$

(the Kronecker delta function $\delta_{\mu\nu}$ is a tensor) so that $\mathcal{T}'_{44} = a_{4\rho} a_{4\sigma} \mathcal{T}_{\rho\sigma}$ if $\mathcal{T}_{\mu\nu}$ transforms as a tensor. If the unprimed frame is the rest frame, where only $\mathcal{T}_{44}$ is non-zero, then

$$\mathcal{T}'_{44} = a_{44}\, a_{44}\, \mathcal{T}_{44} = \frac{\mathcal{T}_{44}}{1 - \frac{v^2}{c^2}} \qquad (4.2.8)$$

Then

$$E' = - \int \mathcal{T}'_{44}\, dx'\, dy'\, dz' = - \int \frac{\mathcal{T}_{44}}{\sqrt{1 - \frac{v^2}{c^2}}}\, dx\, dy\, dz$$

$$= \int \frac{\mathcal{E}}{\sqrt{1 - \frac{v^2}{c^2}}}\, dx\, dy\, dz = \frac{E}{\sqrt{1 - \frac{v^2}{c^2}}} \qquad (4.2.9)$$

The energy of a particle indeed transforms like the fourth component of a four-vector, and the quantity $\mathcal{T}_{\mu\nu}$ transforms as a tensor.

4.3 Possible forms of relativistic gravitational fields

We might therefore guess that $\mathcal{T}_{\mu\nu}$ is the correct source for the gravitational field and write

$$\Box h_{\mu\nu} = k\, \mathcal{T}_{\mu\nu} \qquad (4.3.1)$$

with $h_{\mu\nu}$ having sixteen components. Since $\mathcal{T}_{\mu\nu}$ is symmetric, $h_{\mu\nu}$ will be symmetric, reducing the number of independent components to ten. The conservation laws further reduce the number of independent components : since

$$\frac{\partial \mathcal{T}_{\mu\nu}}{\partial x_\mu} = 0 \qquad (4.3.2)$$

we have

$$\frac{\partial h_{\mu\nu}}{\partial x_\mu} = 0 \qquad (4.3.3)$$

leaving six independent components.

Compare

$$\Box\, h_{\mu\nu} = k\, \mathfrak{J}_{\mu\nu}$$

with

$$\Box\, h = 4\pi G\, \rho = 4\pi G\, \frac{\mathcal{E}}{c^2} = -\frac{4\pi G}{c^2}\, \mathfrak{J}_{44}\ .$$

We must identify h with h_{44} and hence the constant k with $4\pi G/c^2$ and write

$$\Box\, h_{\mu\nu} = -\frac{4\pi G}{c^2}\, \mathfrak{J}^{source}_{\mu\nu}\ . \qquad (4.3.4)$$

The four-scalar interaction, between the field generated by $\mathfrak{J}^{source}_{\mu\nu}$ and a test matter distribution, which reduces to $-h\rho$ in the low velocity limit is (compare the electromagnetic interaction $-\frac{1}{c}\, \mathcal{J}_\mu A_\mu$)

$$-\frac{h_{\mu\nu}\, \mathfrak{J}_{\mu\nu}}{c^2}$$

which for a small test particle of mass m becomes, on integrating over volume,

$$-\frac{h_{\mu\nu}\, T_{\mu\nu}}{c^2} = -\, h_{\mu\nu}\, \frac{\tilde{v}_\mu\, \tilde{v}_\nu}{c^2}\, m \qquad (4.3.5)$$

Suppose we choose to be at rest in the frame of the source of the gravitational field. Then

$$\Box\, h_{44} = -\frac{4\pi G}{c^2}\, \mathfrak{J}^{source}_{44}$$

and because we are at rest in the frame of the source, all other components of $\mathfrak{J}^{source}_{44}$ are zero if the source has no internal motions. These components being zero, they can generate no field and a simple source at rest with respect to the observer only generates an h_{44} component of the gravitational potential. For a time independent field

$$\nabla^2 h_{44} = -\frac{4\pi G}{c^2}\, \mathfrak{J}^{source}_{44}$$

and the solution of this equation gives just the Newtonian potential. The interaction with a test matter distribution is for this case

$$-\frac{h_{\mu\nu}\, \mathfrak{J}_{\mu\nu}}{c^2} = -\frac{h_{44}\, \mathfrak{J}_{44}}{c^2}$$

which is just the Newtonian interaction. Equations (4.3.4) and (4.3.5) thus lead to the Newtonian deflection of light by the sun - we have a problem here, for in this particular case our elaborate relativistic theory gives results in agreement with the naive reasoning in Chapter 1, and in disagreement with experiment.

Nonetheless, before attempting to resolve this discrepancy it is worth noting that effects other than Newtonian come in when elements other than the 44 elements of both $h_{\mu\nu}$ and $\mathfrak{T}_{\mu\nu}$ are non-zero. Just as in electromagnetism, we have differences from the Newtonian case if both source and observed particle are moving with respect to the observer. Even if the centre of mass of the source is at rest, a spinning source will generate extra terms in $h_{\mu\nu}$ giving rise to both a spin-orbit interaction and, if the test particle is also spinning, a spin-spin interaction. This is why the measurement of the precession of an orbiting gyroscope is of interest [1].

We must now examine whether there are any other forms for the source of the gravitational field that reduce to $\mathcal{E}/c^2$ in the low velocity limit. We raised earlier the possibility of a scalar field which would have to couple to a four-scalar: can we construct such a field using $\mathfrak{T}_{\mu\nu}$ as the source of gravitation?

We have already seen that

$$\mathfrak{T}_{44} = a_{44}\, a_{44}\, \mathfrak{T}_{44}(\text{rest frame})$$

so that

$$\mathcal{E} = \frac{\mathcal{E}(\text{rest frame})}{1 - \frac{v^2}{c^2}} .$$

The quantity $\mathcal{E}(1 - v^2/c^2)$ is thus an invariant, the proper energy density, and $\frac{\mathcal{E}}{c^2}(1 - v^2/c^2)$ is the proper mass density. Then

$$E = \int \mathcal{E}\, dx\, dy\, dz = \int \frac{\mathcal{E}\ (\text{rest frame})}{1 - \frac{v^2}{c^2}}\, dx\, dy\, dz$$

$$= \int \frac{\mathcal{E}\ (\text{rest frame})}{\sqrt{1 - \frac{v^2}{c^2}}}\, (dx\, dy\, dz)_{\text{rest frame}}$$

$$= \frac{E\ (\text{rest frame})}{\sqrt{1 - \frac{v^2}{c^2}}}$$

(which just repeats the work leading to equation (4.2.9)).

The four-scalar $\mathcal{E}(1 - v^2/c^2)$ can be formed from $\mathfrak{T}_{\mu\nu}$:

$$\mathcal{E}(1 - v^2/c^2) = -\, \mathfrak{T}_{\mu\mu} \tag{4.3.6}$$

where the quantity $\mathfrak{T}_{\mu\mu}$ is the sum of the diagonal elements of $\mathfrak{T}_{\mu\nu}$, called the trace of $\mathfrak{T}_{\mu\nu}$.

We can thus envisage two forms of scalar interaction between a source density $\mathfrak{J}^{source}_{\mu\nu}$ and a second tensor $\mathfrak{J}_{\mu\nu}$, which reduce to the non-relativistic Newtonian limit. The first gives no results other than Newtonian when the observer is at rest in one or other of the frames of the two masses and is of the form

$$\mathfrak{J}_{\mu\nu}\,\mathfrak{J}^{source}_{\mu\nu}$$

a four-scalar product of two tensors, while the second is of form

$$\mathfrak{J}_{\mu\mu}\,\mathfrak{J}^{source}_{\nu\nu}$$

and is the product of two scalars.

This second term may be written as the product of two tensors, for the Kronecker delta function is a tensor,

$$\mathfrak{J}_{\mu\nu}\,\delta_{\mu\nu}\,\mathfrak{J}^{source}_{\sigma\sigma}$$

and we identify a second kind of possible gravitational field which has a scalar source*. We write the 'Newtonian' field $h^N_{\mu\nu}$ and the scalar field $h^S_{\mu\nu}$ and have

$$\Box\, h^N_{\mu\nu} = k_N\,\mathfrak{J}_{\mu\nu} \tag{4.3.7}$$

and

$$\Box\, h^S_{\mu\nu} = k_S\,\delta_{\mu\nu}\,\mathfrak{J}_{\sigma\sigma}\ , \tag{4.3.8}$$

which is the equation for a scalar field, disguised as a tensor.

If we are in the rest frame of a source with no internal motions, we have only h^N_{44} non-zero, while h^S_{11}, h^S_{22}, h^S_{33} and h^S_{44} are all non-zero. The scalar field yields an interaction which contains a piece dependent on the value of v^2/c^2 for a particle measured with respect to the rest frame of the source, and of course

$$\mathfrak{J}_{\mu\nu}\,h^S_{\mu\nu} \sim \mathfrak{J}_{\mu\mu}\,\mathfrak{J}^S_{\sigma\sigma} \sim \left(1 - \frac{v^2}{c^2}\right)\,\varepsilon\,\mathfrak{J}^S_{44} \tag{4.3.9}$$

in the source frame, so that light is not deflected in the scalar theory. Note also that the scalar field has no off-diagonal terms.

We may make a more general theory in which we form the gravitational potential $h_{\mu\nu}$ out of any suitable combination of $h^N_{\mu\nu}$ and $h^S_{\mu\nu}$. In the low velocity limit we must recover Newtonian gravity, while in the high velocity limit we want twice the Newtonian interaction. Both these features are achieved by setting

---

*This is a directly coupling scalar field. Very different properties may be obtained for an indirectly coupling scalar field [2,3].

$$\Box\, h_{\mu\nu} = -\frac{8\pi G}{c^2}\left\{\mathfrak{J}_{\mu\nu} - \tfrac{1}{2}\,\delta_{\mu\nu}\,\mathfrak{J}_{\sigma\sigma}\right\} \tag{4.3.10}$$

or

$$\Box\, h_{\mu\nu} = -\frac{8\pi G}{c^2}\,\overline{\mathfrak{J}}_{\mu\nu} \ ,$$

where

$$\overline{\mathfrak{J}}_{\mu\nu} = \mathfrak{J}_{\mu\nu} - \tfrac{1}{2}\,\delta_{\mu\nu}\,\mathfrak{J}_{\sigma\sigma} \ .$$

(Note that $\mathfrak{J}_{\sigma\sigma}$ is a scalar and $\delta_{\mu\nu}$ is a tensor.)

These equations express <u>THE</u> tensor theory of gravity, which is distinguished from other theories with tensor fields by a very special characteristic. The field $h_{\mu\nu}$ defined by Eq. (4.3.10) is the classical version of the field which in quantum theory is mediated by the exchange of mass-less, spin 2 particles (gravitons) in much the same way as the vector electromagnetic field is mediated by the exchange of mass-less spin 1 particles, photons. A gravitational wave generated by $\overline{\mathfrak{J}}_{\mu\nu}$ has only two polarisations : this is discussed further in Chapter 7. From now on, when we refer to the tensor theory, we shall mean the theory embodied in Eq. (4.3.10). The scalar theory is embodied in Eq. (4.3.8) although disguised as a tensor field, and the 'Newtonian' theory of Eq. (4.3.7) is part tensor and part scalar : we may loosely call it a half tensor theory. Eq. (4.3.10) can be written

$$\Box\, h_{\mu\nu} = -\frac{8\pi G}{c^2}\left\{\frac{\tilde{v}_\mu \tilde{v}_\nu}{c^2} + \tfrac{1}{2}\,\delta_{\mu\nu}\left(1 - \frac{v^2}{c^2}\right)\right\}\varepsilon \ . \tag{4.3.11}$$

A static field in this tensor theory is thus given by

$$\nabla^2 h_{\mu\nu} = -\frac{8\pi G}{c^2}\left\{\frac{\tilde{v}_\mu \tilde{v}_\nu}{c^2} + \tfrac{1}{2}\,\delta_{\mu\nu}\left(1 - \frac{v^2}{c^2}\right)\right\}\varepsilon \tag{4.3.12}$$

and if $\tilde{v}_1, \tilde{v}_2, \tilde{v}_3 = 0$, $\tilde{v}_4 = ic$, then

$$\begin{aligned} \nabla^2 h_{44} &= \frac{8\pi G}{c^2}\left\{\tfrac{1}{2}\,\varepsilon\right\} \\ \nabla^2 h_{11} &= -\frac{8\pi G}{c^2}\left\{\tfrac{1}{2}\,\varepsilon\right\} \end{aligned} \tag{4.3.13}$$

and similarly for h_{22}, h_{33}. Solving for the fields of a spherically symmetric source of mass M we therefore have

$$h_{44} = -\frac{GM}{r}$$

$$h_{11} = h_{22} = h_{33} = +\frac{GM}{r} \ .$$

The static fields of a mass M can be written for the three cases as

$$h^{S}_{\mu\nu} = \begin{pmatrix} -\frac{GM}{r} & 0 & 0 & 0 \\ 0 & -\frac{GM}{r} & 0 & 0 \\ 0 & 0 & -\frac{GM}{r} & 0 \\ 0 & 0 & 0 & -\frac{GM}{r} \end{pmatrix} \qquad \text{Scalar} \qquad (4.3.14)$$

$$h^{N}_{\mu\nu} = \begin{pmatrix} 0 & 0 & 0 & 0 \\ 0 & 0 & 0 & 0 \\ 0 & 0 & 0 & 0 \\ 0 & 0 & 0 & -\frac{GM}{r} \end{pmatrix} \qquad \text{Newtonian (or 'half tensor')} \qquad (4.3.15)$$

$$h_{\mu\nu} = \begin{pmatrix} \frac{GM}{r} & 0 & 0 & 0 \\ 0 & \frac{GM}{r} & 0 & 0 \\ 0 & 0 & \frac{GM}{r} & 0 \\ 0 & 0 & 0 & -\frac{GM}{r} \end{pmatrix} \qquad \text{Tensor} \qquad (4.3.16)$$

in the rest frame of a spherically symmetric source. In the scalar and tensor cases a fast particle has a gravitational interaction depending on its momentum as well as its energy. In the scalar case this decouples a fast particle, in the tensor case it yields twice the interaction: if we write the interaction with a particle as $-T_{\mu\nu}h_{\mu\nu}/c^2$ we have using the fields of Eqs. (4.3.14) – (4.3.16) :

$$-\frac{T_{\mu\nu}h^{S}_{\mu\nu}}{c^2} = -\left(1-\frac{v^2}{c^2}\right)\frac{GMm}{r} \qquad (4.3.17)$$

$$-\frac{T_{\mu\nu}h^{N}_{\mu\nu}}{c^2} = -\frac{GMm}{r} \qquad (4.3.18)$$

$$-\frac{T_{\mu\nu}h_{\mu\nu}}{c^2} = -\left(1+\frac{v^2}{c^2}\right)\frac{GMm}{r} \;. \qquad (4.3.19)$$

The forces will certainly involve the gradient of these quantities : we expect that the tensor theory will give twice the deflection of light obtaining in our original Newtonian construction.

Thus with the tensor theory we have Lorentz covariant equations connecting the source and the fields, the Newtonian interaction in the low velocity limit and twice the Newtonian interaction for fast particles. We still need to work out in detail the force laws however, and check that this theory yields the proper gravitational redshift for photons despite the extra term in the interaction which seems likely to yield twice the light deflection expected naively. A further problem is whether despite the additional deflection of light we can still maintain that light travels in straight lines in a freely falling box. The next two chapters are devoted to these questions. We must end this chapter by pointing out a probable deficiency in our treatment of gravitation.

4.4   Non-linearity of the field equations

We have set up a theory in which the source of the gravitational field is the energy-momentum density. However, the energy in the gravitational field is itself expected to be a part of this source: strictly speaking we should write

$$\Box h_{\mu\nu} = -\frac{8\pi G}{c^2}\left\{\overline{\mathfrak{I}}_{\mu\nu}\,(\text{gravitational}) + \overline{\mathfrak{I}}_{\mu\nu}\,(\text{everything else})\right\} \qquad (4.4.1)$$

and we expect $\overline{\mathfrak{I}}_{\mu\nu}$ (gravitation) to be composed of bilinear products ('squares') of the derivatives of $h_{\mu\nu}$ itself. Thus for complete consistency we should have a non-linear differential equation for $h_{\mu\nu}$ and the principle of superposition would not hold. This should not be important for weak fields however, and we are mostly concerned with weak fields. Indeed, the linearised (weak field) equations of general relativity can be written in the form (4.3.10).

This kind of problem does not arise in electromagnetism because charge is the source of the electromagnetic field and the electromagnetic field being neutral (photons are uncharged) it does not contribute to its own source.

There are two obvious concrete examples of the difficulties with the linearised theory. First, we expect gravitational waves to be deflected by a gravitational field in the same way as light is deflected. Gravity must therefore be a source of gravity. Secondly, the gravitational energy of a mass M, radius R is $\sim -GM^2/R$ so that the total energy is

$$\sim Mc^2\left(1-\frac{GM}{Rc^2}\right)$$

which goes negative for a finite value of M/R. However, the increasingly negative gravitational energy reduces the gravitational mass and if even for gravitational energy the equivalence between inertial mass and gravitational mass is maintained the total energy can remain positive.

References

[1] See C.W. Misner, K.S. Thorne and J.A. Wheeler, 'Gravitation', (Freeman, 1973) Section 40.7.

[2] R.H. Dicke, Rev. Mod. Phys., 29, 363 (1957).

[3] C.H. Brans and R.H. Dicke, Phys. Rev., 124, 925 (1961).

CHAPTER 5

RELATIVISTIC GRAVITATIONAL FORCES

5.1 The velocity of light

Using the energy-momentum density as the source of relativistic gravitational fields, we have picked out the tensor field with source $\overline{\mathfrak{J}}_{\mu\nu}$. For this case the interaction of a spherically symmetric field due to a source M with a test particle of mass m is given by

$$-\frac{T_{\mu\nu}}{c^2} h_{\mu\nu} = -\left(1+\frac{v^2}{c^2}\right)\frac{GMm}{r} \tag{5.1.1}$$

which reduces in the non-relativistic limit to the familiar Newtonian result. The gradient gives the Newtonian force for $v/c \ll 1$ and twice the Newtonian force for $v/c \to 1$: we may reasonably guess that this form will indeed yield twice the Newtonian deflection for a fast particle - or a light pulse. Before attempting to construct the force laws, it is extremely instructive to examine a necessary consequence of the gravitational deflection of light: the local velocity of light must depend on the local gravitational potential.

We have so far considered the forces acting on relativistic particles, or localised pulses of light. If we consider instead an extended wavefront, deflection of light corresponds to a plane of constant phase, (which is normal to the direction of propagation) being turned through an angle in passing through the gravitational field. This is achieved by a reduction in the phase velocity of the wave as the gravitational field gets stronger (Fig.5.1.1): light passing close to the limb of the Sun will travel more slowly on average than light far from the Sun.

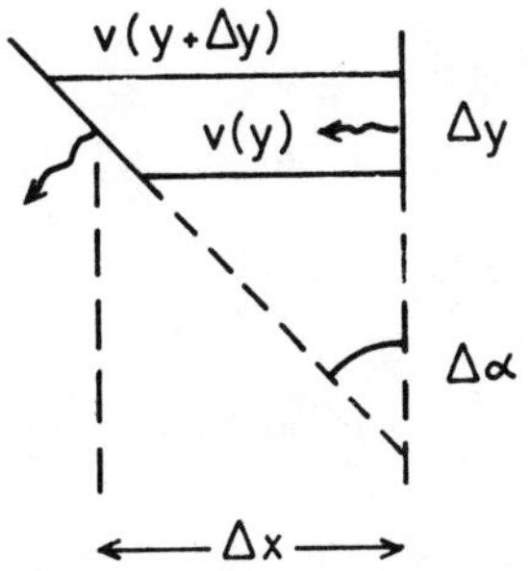

Fig.5.1.1 The normal to a plane of constant phase is rotated through an angle $\Delta\alpha$ in distance Δx as a consequence of the phase velocity varying with y

The gravitational field thus plays the role of a refractive medium and we can calculate an effective refractive index. We set

$$\Delta\alpha = \frac{v(y+\Delta y) - v(y)}{\Delta y}\,\Delta t \qquad \Delta t = \frac{\Delta x}{v} = \frac{n}{c}\,\Delta x$$

and then

$$\Delta\alpha = -\frac{1}{n}\frac{dn}{dy}\Delta x \quad . \tag{5.1.2}$$

This may be identified with the analogous equation obtained from thinking in terms of a relativistic particle :

$$\Delta\alpha = -\frac{dy}{dx} = \left(1+\frac{v^2}{c^2}\right)\frac{GM}{r^3c^2}\frac{\underline{r}\cdot\underline{y}}{y}\Delta x = 2\frac{GM}{r^2c^2}\frac{y}{r}\Delta x \tag{5.1.3}$$

where we are assuming the normal component of the force to be given through the gradient of the tensor interaction. (Fig. 5.1.2 and Chapter 1, section 5.)

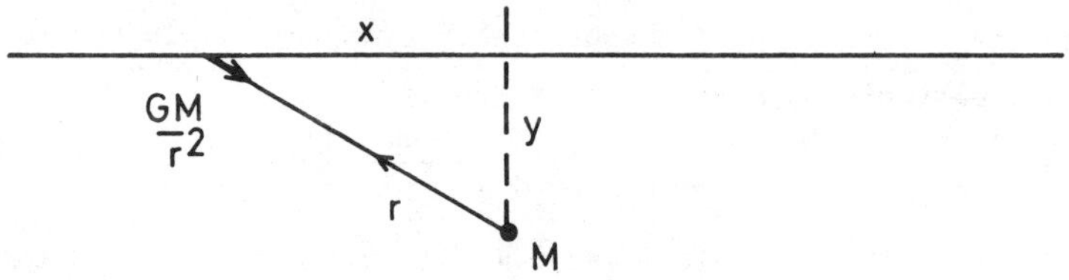

Fig. 5.1.2

We may then make the identification

$$\frac{1}{n}\frac{dn}{dy} = -2\frac{GM}{r^2c^2}\frac{y}{r} \quad . \tag{5.1.4}$$

Since the deflection of a relativistic particle given by the gradient of Eq. (5.1.1) is independent of its energy we expect a photon to be deflected by an amount independent of its frequency. If a pulse of light is deflected by an amount independent of its frequency spectrum, then the gravitational field must be non-dispersive: the effective refractive index must not depend on frequency.

Integrating (5.1.4) in the weak field approximation we find that

$$n(y) - n(\infty) = \int_y^\infty 2\frac{GM}{c^2}\frac{y}{(x^2+y^2)^{\frac{3}{2}}}dy = 2\frac{GM}{rc^2} \quad . \tag{5.1.5}$$

As $r \to \infty$ we want the velocity of light to be c , $n \to 1$ and so with

$$n = 1 + 2\frac{GM}{rc^2} \tag{5.1.6}$$

we achieve consistency between a plane wave picture of light and a pulse (or photon) picture. We can see at once that our full force laws cannot possibly be given by the gradient of (5.1.1) because here we have light slowing down as it enters a gravitational field, despite the tendency of the gradient term to accelerate a particle.

We may at this stage begin to suspect that constructing a relativistic theory of gravity will not be as easy as it was beginning to seem, for the constancy of the velocity of light is fundamental in special relativity.

5.2 Radar ranging in the solar system

It should now be clear that a necessary consequence of the gravitational deflection of electromagnetic waves is a reduction in speed of light in the deflecting field. This is observable and constitutes a further experimental test of relativistic gravitation.

If radar is bounced off a planet when it is near superior conjunction, so that the line of sight passes close to the limb of the Sun, there will be an excess delay introduced in the time for the round trip. The change in path length due to deflection in the Sun's gravitational field is second order in the small quantity $GM_\odot/rc^2$, and so we can compute the time to first order by considering a straight line trajectory (Fig. 5.2.1):

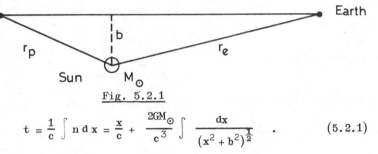

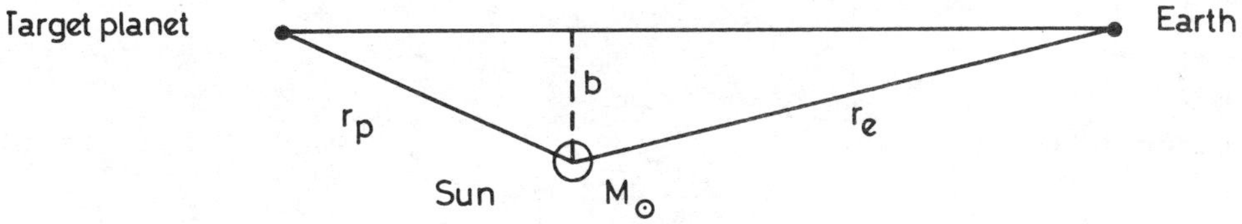

Fig. 5.2.1

$$t = \frac{1}{c}\int n\,dx = \frac{x}{c} + \frac{2GM_\odot}{c^3}\int \frac{dx}{(x^2+b^2)^{\frac{1}{2}}} \quad . \tag{5.2.1}$$

The excess delay is the second term. We evaluate it in pieces. The excess delay introduced in going between Earth and the point of closest approach is

$$\frac{2GM_\odot}{c^3}\left[\ln\left\{x + \sqrt{x^2+b^2}\right\}\right]_0^{\sqrt{r_e^2-b^2}}$$

and in going between the point of closest approach to the Sun and the planet is

$$\frac{2GM_\odot}{c^3}\left[\ln\left\{x + \sqrt{x^2+b^2}\right\}\right]_0^{\sqrt{r_p^2-b^2}} .$$

The one-way journey between Earth and the target planet thus introduces an excess delay

$$\frac{2GM_\odot}{c^3}\ln\left\{\left[\frac{r_e + \sqrt{r_e^2-b^2}}{b}\right]\left[\frac{r_p + \sqrt{r_p^2-b^2}}{b}\right]\right\}$$

and the total excess delay in the round trip is

$$t_E = \frac{4GM_\odot}{c^3} \ln \left\{ \left[\frac{r_e + \sqrt{r_e^2 - b^2}}{b}\right]\left[\frac{r_p + \sqrt{r_p^2 - b^2}}{b}\right] \right\} \tag{5.2.2}$$

with $b \ll r_p, r_e$ we have

$$t_E = \frac{4GM_\odot}{c^3} \ln \left[4 \frac{r_e r_p}{b^2}\right] \tag{5.2.3}$$

and

$$\frac{\partial t_E}{\partial b} = - \frac{8GM_\odot}{bc^3} \quad . \tag{5.2.4}$$

If we take Mercury as the target planet with

$$r_e = 1.495 \times 10^{13} \text{ cm},$$

$$r_p = 0.565 \times 10^{13} \text{ cm},$$

$$b = R_\odot = 7 \times 10^{10} \text{ cm},$$

and

$$M_\odot = 2 \times 10^{33} \text{ gm},$$

we find

$$t_E = 220 \ \mu\text{sec}$$

in a total time of

$$\approx 2 \ \frac{(r_e + r_p)}{c} \approx 22 \text{ mins},$$

(note that 100 μsec is the time taken for light to travel only 30 km) , and

$$\left.\frac{\partial t_E}{\partial b}\right|_{R_\odot} \simeq - 5.7 \times 10^{-16} \text{ s cm}^{-1} .$$

The planets Mercury, Venus and Mars and certain suitable space probes have been used as targets: currently the predictions of general relativity are verified at $\leqslant 3\%$ level [1]. We should note that a measurement of the coefficient in (5.2.4) removes the need to know planetary orbits with great accuracy (the best way to determine them is by radar ranging) and also sweeps under the rug subtleties relating the real earth based system of coordinates, immersed in the Sun's gravitational field, to the idealised coordinates we have used. (See section 9.9.)

As we pointed out in Chapter 1, the refraction of radio waves in the solar corona complicates measurement of both the gravitational deflection of radio waves and of the excess delay introduced by the solar gravitational field in radar ranging. We may now put in a few numbers.

The equation of motion of a free electron in an electric field, neglecting any damping, is :

$$\frac{d^2x}{dt^2} = -\frac{eE}{m} \qquad (5.2.5)$$

and so for a fixed frequency ω

$$x = \frac{eE}{\omega^2 m}$$

and a single oscillating electron corresponds to an oscillating dipole moment

$$-ex = -\frac{e^2E}{m\omega^2} .$$

If there are N electrons per unit volume, the dipole moment per unit volume is

$$-\frac{Ne^2E}{m\omega^2}$$

and the susceptibility of the gas of free electrons is

$$\chi = -\frac{Ne^2}{m\omega^2} .$$

The dielectric constant is thus

$$\epsilon = 1 - \frac{4\pi Ne^2}{m\omega^2} \qquad (5.2.6)$$

and the refractive index

$$n_e = \sqrt{\epsilon} \approx 1 - \frac{2\pi Ne^2}{m\omega^2} \qquad (5.2.7)$$

is to be compared with the refractive index due to gravity

$$n_G = 1 + \frac{2GM_\odot}{rc^2} .$$

In a gravitational field the phase velocity is reduced and the medium is non-dispersive. In a cloud of electrons the phase velocity is increased but the medium is dispersive. A beam of radio waves is deflected away from the Sun by electrons in the corona, but a pulse travels with the group velocity in such a dispersive medium,

$$v_g = c\left(1 - \frac{2\pi Ne^2}{m\omega^2}\right) .$$

We therefore want to compare the quantities $2GM_\odot/rc^2$ and $2\pi Ne^2/m\omega^2$. If we take $N = 5 \times 10^5 \, (r/R_\odot)^{-2}\,\text{cm}^{-3}$, [2], then the ratio

$$\frac{\dfrac{2\pi Ne^2}{m\omega^2}}{\dfrac{2GM_\odot}{rc^2}} = 3.7 \times 10^8 \, \frac{\lambda^2}{r}$$

which is approximately 0.5 for $\lambda = 10\,\text{cm}$, $r = R_\odot$. It should at once be clear why impact parameters smaller than a few solar radii have not been used in these studies: the deflection measurements of reference [3], at $\lambda = 3.7$ cm, employed no impact parameter less than $10 R_\odot$.

5.3 Further remarks on the variable velocity of light

With our tensor theory we have deduced an effective refractive index

$$n = 1 + \frac{2GM}{rc^2} = 1 - \frac{2\varphi}{c^2}$$

for a spherically symmetric field, and found that light slows down as it enters a gravitational field, despite the effect of gradient terms increasing the momentum of a particle. We must search for extra force terms anyway in order to have covariant force laws, but will do well to remember that a velocity v can decrease while a momentum mv increases provided the mass m changes fast enough. (In special relativity with a constant force $dv/dt \to 0$ as $v \to c$ because of the relativistic increase of inertial mass.) We have here a hint that mass may depend on the local gravitational potential, having already found that the velocity of light depends on the local potential.

We may note another interesting result. The dimensionless quantity $2GM/rc^2$ measures the change in the refractive index of empty space in the neighbourhood of a massive body such as the Sun. But all our experiments are done when we are already immersed in the gravitational field of all other bodies in the universe. The potential at the centre of a sphere of constant density ρ is

$$-G\int_0^R \frac{4\pi\rho r^2}{r}\,dr = -2\pi G\rho R^2 \sim -\frac{GM}{R}$$

and so the quantity

$$\frac{2\varphi_{\text{universe}}}{c^2} \approx -\frac{4\pi G\rho_{\text{universe}}R^2_{\text{universe}}}{c^2}.$$

If we take R_{universe} as the Hubble radius, $\approx 10^{28}$ cm and for the mean density of matter, $\rho_{\text{universe}} \approx 10^{-31}\,\text{gm cm}^{-3}$ (still a very uncertain number) then

$$-\frac{2\varphi_{\text{universe}}}{c^2} \approx 10^{-2}$$

which is a number perhaps surprisingly close to unity, considering the large (cosmological) numbers fed into the calculation. Here we have a hint that perhaps our local physics is determined by the properties of the distant parts of the universe [4], through the long range gravitational interaction which, unlike electromagnetic effects, cannot be cancelled out because there are no negative gravitational charges. (See section 10.5.)

5.4 The force laws and equations of motion

In order to determine the equation of motion of a test particle in a gravitational field, we must evaluate the force on a particle, not a force density, and one term must be the generalisation

$$\frac{1}{c^2} T_{\mu\nu} \underline{\nabla} h_{\mu\nu} \rightarrow - m \underline{\nabla} \varphi \qquad \text{in the Newtonian limit.}$$

Then

$$\frac{d\underline{p}}{dt} = \frac{1}{c^2} T_{\mu\nu} \underline{\nabla} h_{\mu\nu} \qquad + \text{ other terms} \tag{5.4.1}$$

and more generally the time derivative of the momentum-energy four-vector p_ρ will be given by

$$\frac{dp_\rho}{dt} = \frac{1}{c^2} T_{\mu\nu} \frac{\partial}{\partial x_\rho} h_{\mu\nu} \qquad + \text{ other terms.} \tag{5.4.2}$$

The other terms must presumably be of form

$$\frac{1}{c^2} T_{\mu\nu} \frac{\partial}{\partial x_\mu} h_{\rho\nu} ,$$

noting that the only other possible force field to couple to $T_{\mu\nu}$ is $\dfrac{\partial h_{\mu\rho}}{\partial x_\nu}$ which gives nothing new because $T_{\mu\nu}$ and $h_{\mu\nu}$ are symmetric. Then we must have

$$\frac{dp_\rho}{dt} = \frac{1}{c^2} T_{\mu\nu} \frac{\partial h_{\mu\nu}}{\partial x_\rho} + \frac{\alpha}{c^2} T_{\mu\nu} \frac{\partial h_{\rho\nu}}{\partial x_\mu} \tag{5.4.3}$$

and we have to determine the quantity α in order to establish the equations of motion.

Consider the spherically symmetric fields generated by a source M, the fields not being explicitly time dependent. Then

$$h_{11} = h_{22} = h_{33} = - h_{44} = - \varphi = \frac{GM}{r}$$

and we can write, setting $c = 1$ for convenience

$$\frac{dp_\rho}{dt} = m\tilde{v}_\mu \tilde{v}_\nu \frac{\partial h_{\mu\nu}}{\partial x_\rho} + \alpha m \tilde{v}_\mu \tilde{v}_\nu \frac{\partial h_{\rho\nu}}{\partial x_\mu} \tag{5.4.4}$$

where

$$\tilde{v}_i = v_i = \frac{dx_i}{dt} , \quad \tilde{v}_4 = i$$

and the quantities $\tilde{v}_\mu$ do not make up a four-vector. Then since $\partial\varphi/\partial t = 0$

$$\frac{d\underline{p}}{dt} = - m(1 + v^2)\,\underline{\nabla}\varphi - \alpha m\,\underline{v}\,\underline{v}\,.\,\underline{\nabla}\varphi \tag{5.4.5a}$$

$$\frac{dp_4}{dt} = i\,\alpha m\,\underline{v}\,.\,\underline{\nabla}\varphi \;\;,\quad \frac{dE}{dt} = \alpha m\,\underline{v}\,.\,\underline{\nabla}\varphi\;. \tag{5.4.5b}$$

It is possible to write these equations in terms of a single scalar potential φ only because we are considering the particularly simple case of a spherically symmetric source at rest in the reference frame. It must always be remembered that the gravitational potential is a (16 component) second rank tensor.

The rate of change of kinetic energy is the scalar product of force and velocity

$$\frac{dE}{dt} = \underline{v}\,.\,\frac{d\underline{p}}{dt}\;.$$

This is familiar in classical mechanics: it also holds in relativistic mechanics since $E^2 - p^2 = m_0^2$ (in units with $c = 1$) whence

$$E\,\frac{dE}{dt} = \underline{p}\,.\,\frac{d\underline{p}}{dt}\;.$$

In order to satisfy this relation we must set $\alpha = -1$ when (5.4.5a) and (b) become

$$\frac{d\underline{p}}{dt} = - m(1 + v^2)\,\underline{\nabla}\varphi + m\,\underline{v}\,\underline{v}\,.\,\underline{\nabla}\varphi \tag{5.4.6a}$$

$$\frac{dE}{dt} = - m\,\underline{v}\,.\,\underline{\nabla}\varphi\;. \tag{5.4.6b}$$

We may note that (5.4.6b) is unchanged in the Newtonian limit, while (5.4.6a) reduces to the Newtonian result as $v \to 0$. Rewriting (5.4.6a) as

$$\frac{d\underline{p}}{dt} = - m\,\underline{\nabla}\varphi - m\left\{v^2\,\underline{\nabla}\varphi - \underline{v}\,\underline{v}\,.\,\underline{\nabla}\varphi\right\}$$

we see that the second term can be written as

$$m\,\underline{v} \times (\underline{v} \times \underline{\nabla}\varphi)$$

and is always at right angles to the motion. Such a force does no work.

The term in $\underline{v}\,\underline{v}\,.\,\underline{\nabla}\varphi$ has no component at right angles to the direction of motion. The only force term with a component at right angles to the direction of motion is

$$- m(1 + v^2)\,\underline{\nabla}\varphi$$

and so we do indeed get twice the Newtonian deflection of light, on applying the treatment given in section 1.5.

We may thus infer that (5.4.4) should be written

$$\frac{dp_\rho}{dt} = m\tilde{v}_\mu\tilde{v}_\nu \frac{\partial h_{\mu\nu}}{\partial x_\rho} - m\tilde{v}_\mu\tilde{v}_\nu \frac{\partial h_{\rho\nu}}{\partial x_\mu} \qquad (5.4.7)$$

for the case of a general gravitational field. However, while p_ρ is a four-vector, it is differentiated with respect to time, and on the right hand side we have

$$\tilde{v}_\mu\tilde{v}_\nu = \frac{dx_\mu}{dt}\frac{dx_\nu}{dt} .$$

Thus although this equation is attractive, we must see if it is indeed Lorentz covariant. We may write

$$\frac{dp_\rho}{dt} = \frac{dp_\rho}{d\tau}\frac{d\tau}{dt} .$$

The proper time τ is an invariant and so $dp_\rho/d\tau$ is a four-vector. With

$$\frac{dx_\mu}{dt} = \frac{dx_\mu}{d\tau}\frac{d\tau}{dt}$$

we have

$$\frac{dp_\rho}{d\tau} = m\frac{d\tau}{dt}\frac{dx_\mu}{d\tau}\frac{dx_\nu}{d\tau}\left\{\frac{\partial h_{\mu\nu}}{\partial x_\rho} - \frac{\partial h_{\rho\nu}}{\partial x_\mu}\right\} .$$

The quantity m is the inertial mass of the particle and

$$\Delta t = \Delta\tau(1-v^2)^{-\frac{1}{2}}$$

so

$$m\frac{d\tau}{dt} = m(1-v^2)^{\frac{1}{2}} = m_0 .$$

We can therefore write (5.4.7) in the form

$$\frac{dp_\rho}{d\tau} = m_0\frac{dx_\mu}{d\tau}\frac{dx_\nu}{d\tau}\left\{\frac{\partial h_{\mu\nu}}{\partial x_\rho} - \frac{\partial h_{\rho\nu}}{\partial x_\mu}\right\} \qquad (5.4.7a)$$

which is indeed manifestly covariant.

We may now calculate the acceleration of a particle in a gravitational field. We have the equations (5.4.6) which give the rate of change of energy and momentum and must combine them noting that because mass increases with velocity the rate of change of velocity will be less than the rate of change of momentum. Given

$$\underline{p} = E\underline{v} = m\underline{v} \qquad (5.4.8)$$

$$\frac{d\underline{v}}{dt} = \frac{1}{E}\frac{d\underline{p}}{dt} - \frac{\underline{v}}{E}\frac{dE}{dt} = -(1+v^2)\,\underline{\nabla}\varphi + 2\underline{v}\,\underline{v}\cdot\underline{\nabla}\varphi$$

and

$$\underline{v}\cdot\frac{d\underline{v}}{dt} = -\underline{v}\cdot\underline{\nabla}\varphi + v^2\,\underline{v}\cdot\underline{\nabla}\varphi \to 0 \qquad (5.4.9)$$

as $v \to 1$.

But in the previous section we found that a consistent description of the propagation of light in a gravitational field could only be obtained with

$$c_\varphi = \frac{c}{n} = c\,(1 + 2\varphi)$$

or $c_\varphi = 1 + 2\varphi$ with $c = 1$ far away from the source. Then

$$\frac{d}{dt}\left(c_\varphi^2\right) = 2\,\underline{c}_\varphi \cdot \frac{d\underline{c}_\varphi}{dt} = 4\,\frac{d\varphi}{dt}$$

with $c_\varphi \simeq 1$. Since $\dfrac{\partial\varphi}{\partial t} = 0$, $\dfrac{d\varphi}{dt} = \underline{c}_\varphi \cdot \underline{\nabla}\varphi$

so

$$\underline{c}_\varphi \cdot \frac{d\underline{c}_\varphi}{dt} = 2\,\underline{c}_\varphi \cdot \underline{\nabla}\varphi \tag{5.4.10}$$

and this result should also hold for a particle in the limit $v \to c_\varphi$. Eqs. (5.4.9) and (5.4.10) are thus inconsistent. We must examine the assumptions used in extracting (5.4.9).

Equation (5.4.9) was constructed from Eq. (5.4.7), using $\underline{p} = E\underline{v}$. Eq. (5.4.7) in turn depended on the identification $\alpha = -1$. This was obtained by insisting that

$$E\,\frac{dE}{dt} = \underline{p} \cdot \frac{d\underline{p}}{dt}$$

which in turn depended on the relation

$$E^2 - p^2c^2 = m_0^2\,c^4 .$$

In special relativity the relation $\underline{p} = E\underline{v}$ yields

$$E = \frac{m_0\,c^2}{\sqrt{1 - \dfrac{v^2}{c^2}}} . \tag{5.4.11}$$

This relation gives us the clue to the source of our inconsistency, for Eq. (5.4.11) implies $v \leq c$. In a gravitational field we find the local velocity of light is reduced, and we have no hope of finding the same laws of physics in all freely falling frames if particles can outrun light in a gravitational potential. If we substitute for (5.4.11) the relation

$$E = \frac{m_0\,c^2}{\sqrt{1 - \dfrac{n^2v^2}{c^2}}} \tag{5.4.12}$$

the maximum particle velocity is then $c_\varphi = \dfrac{c}{n}$. Keeping

$$E^2 - p^2\,c^2 = m_0^2\,c^4$$

we obtain (5.4.12) by setting

$$\underline{p} = n\,E\,\underline{v} = (1 - 2\varphi)\,E\,\underline{v} \tag{5.4.13}$$

and here we are departing from familiar special relativity. Then

$$n E \frac{d\underline{v}}{dt} = \frac{d\underline{p}}{dt} - \underline{v}\, n \frac{dE}{dt} - E \underline{v} \frac{dn}{dt}$$

so that to first order in the potential φ we have

$$\frac{d\underline{v}}{dt} = -(1+v^2)\,\underline{\nabla}\varphi + 4\underline{v}\ \underline{v}\cdot\underline{\nabla}\varphi \qquad (5.4.14)$$

$$\underline{v}\cdot\frac{d\underline{v}}{dt} = -\underline{v}\cdot\underline{\nabla}\varphi + 3v^2\ \underline{v}\cdot\underline{\nabla}\varphi$$

which indeed gives the correct deceleration as $v \to 1$ and we have at least established consistency with Eq. (5.4.10), to first order in φ, and have the equations

$$\frac{d\underline{p}}{dt} = -E(1+v^2)\,\underline{\nabla}\varphi + E\underline{v}\ \underline{v}\cdot\underline{\nabla}\varphi$$

$$\frac{dE}{dt} = -E\underline{v}\cdot\underline{\nabla}\varphi \qquad (5.4.15)$$

together with new kinematic equations

$$\underline{p} = n E \underline{v} \qquad E^2 - p^2 = m_0^2 \qquad (5.4.16)$$

which may be written

$$E = \frac{m_0}{\sqrt{1-n^2v^2}} \qquad \underline{p} = \frac{m_0 n \underline{v}}{\sqrt{1-n^2v^2}} \qquad (5.4.17)$$

The second of Eqs. (5.4.17) shows that the quantity $n\,m_0$ is playing the role of the inertial mass. The interpretation of these equations is however a little more complicated and it is instructive to consider the effect of the gravitational potential on the physics of a small region of space.

5.5 The effects of local forces

Consider a particle in a gravitational field which is acted on by some arbitrary force $\underline{F}$, for example an electron in an atom. The equations of motion are

$$\frac{d\underline{p}}{dt} = -E(1+v^2)\,\underline{\nabla}\varphi + E\underline{v}\ \underline{v}\cdot\underline{\nabla}\varphi + \underline{F} \qquad (5.5.1)$$

$$\frac{dE}{dt} = -E\underline{v}\cdot\underline{\nabla}\varphi + \underline{v}\cdot\underline{F}\ . \qquad (5.5.2)$$

If $\underline{F} = 0$ then to first order in φ

$$\frac{d}{dt}\left\{E(1+\varphi)\right\} = \frac{dE}{dt} + E\underline{v}\cdot\underline{\nabla}\varphi$$

The quantity $E(1+\varphi)$ is then conserved in free fall. Similarly, to first order in φ:

$$\frac{d\underline{p}}{dt} - E\underline{v}\,\underline{v}\cdot\underline{\nabla}\varphi = \frac{d}{dt}\left\{\underline{p}\,(1-\varphi)\right\}$$

since

$$\frac{d\varphi}{dt} = \frac{\partial\varphi}{\partial t} + \underline{v}\cdot\underline{\nabla}\varphi .$$

This suggests that when a force $\underline{F}$ is acting we should write

$$\frac{d}{dt}\left\{E\,(1+\varphi)\right\} = \underline{v}\cdot\underline{F} . \qquad (5.5.3)$$

The quantity $E(1+\varphi)$ is then acting as the energy in so far as the response to local fields is concerned. In the absence of local fields other than gravitation (that is, in free fall) this quantity is conserved, corresponding of course to the compensation of potential and kinetic energy.

This identification tells us that if we lower a system gently on a string into a gravitational field, its energy goes as $E \to E(1+\varphi)$: this is the identification we made to calculate the redshift for Newtonian gravity. We therefore write the energy of a particle of velocity v in a potential φ as

$$E \to E_{\varphi} = E(1+\varphi) = \frac{m_o(1+\varphi)}{\sqrt{1-n^2v^2}} . \qquad (5.5.4)$$

Similar reasoning suggests that we may also write

$$\frac{d}{dt}\left\{\underline{p}(1-\varphi)\right\} = -\,E(1+v^2)\,\underline{\nabla}\varphi + \underline{F}$$

and if F is large in comparison with the gravitational force the quantity $\underline{p}_{\varphi} = \underline{p}(1-\varphi)$ would then act as the momentum of the particle: steadily lowering an atomic system into a gravitational potential changes its response to the force and we set

$$\underline{p} \to \underline{p}_{\varphi} = \underline{p}(1-\varphi) = \frac{(1-\varphi)\,n\underline{v}\,m_o}{\sqrt{1-n^2v^2}} = \frac{(1-3\varphi)\,m_o\underline{v}}{\sqrt{1-n^2v^2}} . \qquad (5.5.5)$$

We may interpret this as giving us a rest mass which is dependent on the local gravitational potential

$$m_o \to m_{o\varphi} = m_o(1-3\varphi) .$$

Since φ is negative, mass increases in a gravitational field and indeed a particle may be slowed down despite its increasing momentum.

With these new definitions of energy, momentum and rest mass, we find

$$m_{o\varphi}\,c_{\varphi}^2 = m_o(1-3\varphi)\,c^2(1+4\varphi) = m_o\,c^2(1+\varphi)$$

to be compared with $E_\varphi = E(1+\varphi)$: the quantity $m_{0_\varphi} c_\varphi^2$ is acting as the rest mass energy. Further

$$E_\varphi^2 - p_\varphi^2 c_\varphi^2 = m_{0_\varphi}^2 c_\varphi^4 \tag{5.5.6}$$

and of course

$$\underline{p}_\varphi = \frac{m_{0_\varphi} \underline{v}}{\sqrt{1 - v^2/c_\varphi^2}} \tag{5.5.7}$$

$$E_\varphi = \frac{m_{0_\varphi} c_\varphi^2}{\sqrt{1 - v^2/c_\varphi^2}} \quad . \tag{5.5.8}$$

Thus all the kinematic formulae of special relativity emerge as being locally true in a region of space where φ does not change perceptibly.

We can also write the free-fall equations as

$$\frac{d\underline{p}_\varphi}{dt} = - E(1+v^2)\, \underline{\nabla}\varphi = - E_\varphi(1+v^2)\, \underline{\nabla}\varphi \,,$$

$$\frac{dE_\varphi}{dt} = 0$$

to first order in φ which with

$$\underline{p}_\varphi = \frac{E_\varphi \underline{v}}{c_\varphi^2}$$

yields again

$$\frac{d\underline{v}}{dt} = - (1+v^2)\, \underline{\nabla}\varphi + 4\underline{v}\; \underline{v}\cdot\underline{\nabla}\varphi$$

where the last term comes from the variation of c_φ with the potential φ.

The identification $m_{0_\varphi} = (1 - 3\varphi)\, m_0$ implies the existence of acceleration dependent forces in a gravitational field, which we have not built in. We return to this point in section 6.5: we may wonder whether inertial mass is in fact due to the gravitational potential of the universe (Mach's principle) [4] .

5.6 Gravitational deflection and gravitational redshift

We have now reconciled the apparently discordant observations that the gravitational redshift is found to be in accord with elementary reasoning, while the gravitational deflection of light is twice the naive value. The force acting on a particle has a relativistic term of order $(v/c)^2$, one piece of which acts in the direction of $-\underline{\nabla}\varphi$ and the other piece of which acts along the path of the particle. The gravitational deflection is due to that component of the force which is normal to the path and this is multiplied by a factor $(1+\frac{v^2}{c^2})$, giving the required factor of 2 in the deflection of light. An unwanted factor in the energy is avoided because the two pieces of the relativistic correction taken together always act at right angles to the motion and so do not affect the energy.

If we adopt a wave picture for light in our flat space coordinates we must argue in the following way. The gravitational potential acts as a dielectric, and in a time independent gravitational field the dielectric constant is not explicitly time dependent. Consequently the frequency is a constant. The frequency with which the light is emitted must therefore be shifted by $(1+\varphi)$. For a concrete example, suppose we lower an electron and positron into the potential. The rest energy of the pair changes by $(1+\varphi)$. Setting $h\nu = mc^2$ we see that the annihilation photons are shifted in energy and frequency by $(1+\varphi)$ with respect to annihilation photons produced outside the potential, and do not change their total energy in falling freely out of the gravitational field.

The frequency of radiation produced classically reflects the frequency of the oscillator. We must therefore find that the frequency of an oscillator in a gravitational field is reduced by the gravitational potential: for a consistent picture clocks run slower in a gravitational potential. In both the tensor and scalar theories we have discussed, the mechanism which arranges this also affects the lengths of measuring sticks: gravitational potentials distort the instruments we use in surveying reference frames. This is the subject of the next chapter.

References

[1] I.I. Shapiro, et al., Phys. Rev. Lett., 26, 1132 (1971).
J.D. Anderson et al., Ap. J., 200, 221 (1975).

[2] C.C. Counselman and J.M. Rankin, Ap. J., 185, 357 (1973).

[3] C.C. Counselman et al., Phys. Rev. Lett., 33, 1621 (1974).

[4] See R.H. Dicke, 'The Theoretical Significance of Experimental Relativity', (Gordon & Breach, 1965).

CHAPTER 6
THE DISTORTION OF REFERENCE FRAMES

6.1 Introduction

In the last chapter we found that a gravitational potential changes the velocity of light so that

$$c \to c_\varphi = c(1 + 2\varphi)$$

and also changes the rest energy of an object

$$E \to E_\varphi = E(1 + \varphi)$$

the latter relation implying that the frequency of a periodic motion is changed by a gravitational field

$$\omega \to \omega(1 + \varphi)$$

and hence that clocks are slowed down.

If we wish to elevate the principle of equivalence of inertial and gravitational mass (perhaps better called the principle of unique gravitational acceleration) to a principle which states that

> " In all localised freely falling frames the laws of physics are the same, Lorentz covariant and contain the same numerical constants ",

then we must find the same numerical value for the velocity of light in any freely falling local frame. Far away from the source of the gravitational field we use a clock to time light over a path in the laboratory measured with a measuring stick. The whole laboratory is now moved deep into the gravitational field and the measurement repeated. With respect to the reference frame far away from the source, light has been slowed down,

$$c \to c_\varphi = c(1 + 2\varphi)$$

and the clock has been slowed down but by a smaller amount $\omega \to \omega(1 + \varphi)$. The velocity of light is determined by the ratio of the length of the measuring stick to the number of ticks N of the clock while light is travelling from one end to the other, $N \sim \omega^{-1}$

$$c = \frac{L}{N} \to \frac{L_\varphi}{N_\varphi} = \frac{L_\varphi}{N}(1 + \varphi) \ .$$

If

$$L \to L_\varphi = L(1 + \varphi)$$

then

$$\frac{L_\varphi}{N_\varphi} = \frac{L}{N}(1 + 2\varphi)$$

and the measuring stick is contracted so that a measurement of c_φ using standard measuring sticks and clocks immersed in the same field will yield the same numerical result for the velocity of light as an identical experiment performed far from the source of the potential φ.

An alternative way of stating the same result is the following. We have inferred that atomic clocks will be slowed down, $\omega \to \omega(1+\varphi)$. A clock could also be constructed by bouncing a pulse of light backwards and forwards between two mirrors of known separation. If the laws of physics are the same in any local freely falling frame, this clock must maintain synchronism with an atomic clock as the pair fall into a gravitational potential. Light is slowed down, $c \to c(1+2\varphi)$ and so the light clock will be slowed down by this amount $\omega_L \to \omega_L(1+2\varphi)$ unless the structure linking the mirrors shrinks. If $L \to L(1+\varphi)$ then $\omega_L \to \omega_L(1+\varphi)$ and the atomic and light clocks measure the same local elapsed time.

We must therefore examine the physics of atoms in a gravitational potential. We expect to find that atomic frequencies and dimensions are both changed by a factor $(1+\varphi)$.

6.2 Atoms in gravitational fields: the change of scale

We consider an atom deep in a potential φ and ignore the gravitational force (which we can always do by having the atom in free fall) and the tidal forces which for an atom should be negligible. We then have for the momentum of a particle of mass m_o

$$\underline{p}_\varphi = \frac{m_o(1-3\varphi)\,\underline{v}}{\sqrt{1-n^2v^2}}$$

and $m_\varphi = (1-3\varphi)\,m$. The change of mass alone is sufficient to change the atomic structure in the presence of a gravitational potential, but it cannot be the only effect.

It is adequate for our purposes to use the simple Bohr model of the atom. The equations governing atomic structure are

$$\frac{mv^2}{r} = \frac{e^2}{r} \qquad E = \tfrac{1}{2}mv^2 - \frac{e^2}{r} \tag{6.2.1}$$

and the Bohr condition

$$m\,v\,r = \hbar\,. \tag{6.2.2}$$

(If we were constructing a classical model, (6.2.2) would be replaced by $m\,v\,r$ = constant, conservation of angular momentum.)

The characteristic dimension of the atom, the Bohr radius, is given by

$$r_B = \frac{\hbar^2}{m e^2} \tag{6.2.3}$$

and

$$E \sim \frac{me^4}{\hbar^2} \quad , \quad \omega \sim \frac{me^4}{\hbar^3} \; . \tag{6.2.4}$$

We expect the electric field that holds the atom together to be modified by the gravitational potential, for the electromagnetic energy-momentum tensor is coupled to the potential $h_{\mu\nu}$. The effect of this coupling on electromagnetic waves could be represented by an effective refractive index $n = 1-2\varphi$. We also know that the electrostatic energy of a system varies as $(1+\varphi)$ from the Eötvös-Dicke experiments. The energy density in an electric field is

$$\frac{\epsilon E^2}{8\pi} \; , \tag{6.2.5}$$

and

$$\underline{\nabla} \cdot (\epsilon \underline{E}) = 4\pi\rho \tag{6.2.6}$$

in classical electromagnetic theory. Then

$$|E| \sim \frac{e}{\epsilon r^2}$$

and the energy $\mathcal{E}$ stored in a system of dimension r is

$$\mathcal{E} \sim \frac{e^2}{\epsilon r} \; .$$

Such a system could be a capacitor, or perhaps an atomic nucleus. If we let ϵ take up the effect of the gravitational potential on a given field then the effect of ϵ together with the change in r caused by the gravitational potential must cumulatively change $\mathcal{E}$ by a factor $(1+\varphi)$. If therefore the dimensions of a physical object vary as $(1+\varphi)$, as we have guessed, then

$$\epsilon = 1-2\varphi \, . \tag{6.2.7}$$

Substitute this factor in the equations describing the atom by replacing e^2 by e^2/ϵ in Eqs. (6.2.1) and find

$$r_B \to \frac{\hbar^2 \epsilon}{m e^2} \, .$$

We also must replace m by $m(1-3\varphi)$ to find the dimensions of the electron orbit in the potential φ and find at once

$$r_B \to r_B (1+\varphi)$$

as we expected. We also have

$$E\, , \omega \sim \frac{m_\varphi e^4}{\epsilon^2} \sim (1+\varphi)\, m e^4$$

and so have achieved an entirely consistent picture.

We may also note that with magnetic energy density given by

$$\frac{\mu H^2}{8\pi} \tag{6.2.8}$$

and

$$\underline{\nabla} \times \underline{H} = \frac{4\pi}{c} \underline{\mathcal{J}}$$

where c is the velocity of light far from the source of the gravitational field,

$$|H| \sim e \frac{v}{r^2}$$

and the magnetic energy $\mathfrak{m}$ of a system varies as

$$\mathfrak{m} \sim \mu \frac{e^2 v^2}{r} \quad . \tag{6.2.9}$$

With $r \sim (1+\varphi)$ and $v \sim \frac{r}{t} \sim (1+2\varphi)$, then $\mu \sim (1-2\varphi)$ if the magnetic energy of a real physical system is also to vary as $(1+\varphi)$. We therefore find that we have a completely consistent picture if we set $\epsilon = \mu = 1-2\varphi$ and this moreover yields the effective refractive index we require for electromagnetic waves, through the relation

$$n = \sqrt{\epsilon\mu} \quad . \tag{6.2.10}$$

The quantities ϵ and μ play their usual roles for both static fields and radiation, and an atomic system is indeed shrunk and slowed down as it is lowered into a spherically symmetric gravitational field. A capacitor shrinks and its electric field is diluted so that $\mathcal{E} \sim (1+\varphi)$ and we infer from the Eötvös-Dicke experiments that a nucleus, held together by forces we have not considered, shrinks in the same way.

In our original attempts to calculate the deflection of light in a gravitational field we followed two approaches which gave the same results. We first adopted as an axiom the principle that light travels in straight lines in a freely falling reference frame. Secondly we evaluated the deflection by calculating the change of momentum due to a Newtonian force. These two approaches gave the same answer, wrong by a factor of 2. We can now see that with the picture of gravitation we have developed light is deflected by twice the Newtonian amount and yet still travels in straight lines across a freely falling box. A system of measuring rods is distorted in a gravitational potential so as to remove that piece of the curvature of light that is not removed by the gravitational acceleration. An observer located in such a reference frame has however no way of detecting this distortion.

In a distance Δx the gravitational deflection in the tensor theory is

$$\Delta\alpha = \frac{2GM}{r^2c^2}\,\Delta x\,. \qquad (6.2.11)$$

Half of this disappears if the reference frame is falling freely.

Now the deflection (6.2.11) is measured with respect to an undistorted reference frame a large distance from the source of the gravitational potential. Consider a box which when surveyed in this Olympian reference frame is cubical. Take the same box and immerse it in the gravitational potential. From the point of view of the Olympian observer, the bottom is shrunk more than the top, but the local observer in the box has no way of telling that his certified rectangular frame is bent like a banana. He counts the same number of atomic diameters at the top as at the bottom, and he finds light takes the same time (as measured by a local clock) to cross the bottom as to cross the top. If the box is of side ΔL in the Olympian frame then the top is at a potential φ_T and is of side

$$\Delta L\left(1+\frac{\varphi_T}{c^2}\right)$$

and the bottom is of side

$$\Delta L\left(1+\frac{\varphi_B}{c^2}\right)\,.$$

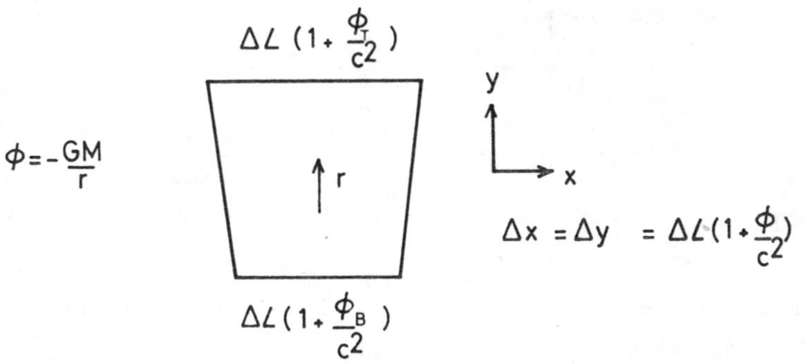

Fig.6.2.1 The distortion of a nominally rectangular box by the differential shrinking of top and bottom in a gravitational field

The angle between the walls is thus

$$\frac{1}{c^2}\,\Delta L\,\frac{\Delta\varphi}{\Delta y} \rightarrow \frac{1}{c^2}\,\Delta x\,\frac{\partial\varphi}{\partial r} = \frac{GM}{r^2c^2}\,\Delta x\,.$$

If light is launched at 90° to one wall in a freely falling frame it impinges on the other wall at 90°. The local observer thus claims that it has travelled through his reference system in a straight line. The Olympian observer agrees, and constructs the gravitational deflection of light to be observed at large distances from the acceleration of the observer plus the distortion of the local frame.

If the local frame is not freely falling, but is at rest, the local observer will see a deflection

$$\Delta\alpha = \Delta x \frac{GM}{r^2 c^2} \tag{6.2.12}$$

and will observe a reading GM/r^2 on an accelerometer. These observations are in accord with the observations that would be made in a space vehicle with the same accelerometer reading. It is thus the distortion of measuring rods by a gravitational potential that has introduced a new feature and has allowed locally equivalent physics in a gravitational field and a space vehicle and yet twice the originally expected deflection of light. The flat space coordinate system we have been working in is perfectly well defined, but is unobservable in the context of present day physics: we would need something like the hyperwaves of science fiction to detect by local measurements the distortion of standard rods and clocks.

We may examine what would happen with either the scalar theory or the Newtonian theory by writing for a spherical field

$$h_{\mu\nu} = \begin{pmatrix} a\frac{GM}{r} & 0 & 0 & 0 \\ 0 & a\frac{GM}{r} & 0 & 0 \\ 0 & 0 & a\frac{GM}{r} & 0 \\ 0 & 0 & 0 & -\frac{GM}{r} \end{pmatrix} \tag{6.2.13}$$

$$\varphi = -\frac{GM}{r}$$

where $a = +1$ for the tensor theory, $a = -1$ for the scalar theory and $a = 0$ for the Newtonian theory. Then Eqs. (5.4.5) become

$$\left.\begin{aligned} \frac{d\underline{p}}{dt} &= -m\,(1 + a^2v^2)\,\underline{\nabla}\varphi - \alpha a\, m\underline{v}\,\underline{v}\cdot\underline{\nabla}\varphi \\ \frac{dE}{dt} &= \alpha m\,\underline{v}\cdot\underline{\nabla}\varphi \end{aligned}\right\}. \tag{6.2.14}$$

In order to satisfy the relation

$$E\frac{dE}{dt} = \underline{p}\cdot\frac{d\underline{p}}{dt}$$

we must set $\alpha = -1$ when we have

$$\frac{d\underline{p}}{dt} = -m(1 + a v^2)\ \underline{\nabla}\varphi + a m \underline{v}\ \underline{v} \cdot \underline{\nabla}\varphi$$

$$\frac{dE}{dt} = -m\ \underline{v} \cdot \underline{\nabla}\varphi\ .$$

In a scalar theory there is no deflection of light because $a = -1$. We may therefore set $\underline{p} = E\underline{v}$ (i.e. $\underline{p} = nE\underline{v}$ with $n = 1$) and find no acceleration as $v \to c$.

In the scalar theory we therefore have

$$E^S_\varphi = E(1 + \varphi)$$

$$p^S_\varphi = \underline{p}(1 - a\varphi) = \underline{p}(1 + \varphi)$$

$$c_\varphi = c\ .$$

The mass therefore goes also as $(1 + \varphi)$, and for an atom

$$r_B \sim \frac{1}{m e^2}\quad ,\quad E, \omega \sim m e^4\ .$$

Since light is not deflected the effective dielectric constant is plausibly 1 and so $r_B \sim (1 - \varphi)$ and $E, \omega \sim (1 + \varphi)$. That is, lengths are expanded and clocks are slowed down. We clearly get the same gravitational redshift.

Light again goes straight across a freely falling box, as seen by an observer in the box, because although from the point of view of an Olympian observer light is not deflected, the bottom of the box is expanded more than the top, inducing the opposite effect from that obtaining with the tensor field.

In the Newtonian case, we have $a = 0$. We require $n = 1 - \varphi$ and set

$$\underline{p} = nE\underline{v} = (1 - \varphi)\, E\underline{v}\ .$$

Then

$$\frac{d\underline{v}}{dt} = -\underline{\nabla}\varphi + 2\underline{v}\ \underline{v} \cdot \underline{\nabla}\varphi$$

and as $v \to 1$

$$\underline{v} \cdot \frac{d\underline{v}}{dt} \to v^2\ \underline{v} \cdot \underline{\nabla}\varphi$$

which is the correct deceleration.

We then identify

$$E^N_\varphi = E(1 + \varphi) \quad \text{as usual}$$

$$p^N_\varphi = (1 - a\varphi)\, nE\underline{v} = (1 - \varphi)\,\underline{p} \quad \text{and} \quad m^N_{0\,\varphi} = (1 - \varphi)\, m_0$$

$$c^N_\varphi = c(1 + \varphi)$$

With $\epsilon = (1-\varphi)$ and $m = m(1-\varphi)$ the Bohr radius r_B is unchanged. Once more $E, \omega \sim me^4 \sim m\epsilon^{-2} \sim 1+\varphi$. In the Newtonian theory (an equal mixture of tensor and scalar) there is no distortion of measuring rods and the elementary reasoning of Chapter 1 survives. Clocks, of course, are still affected.

6.3 The weight of an atom

We have been working with the fields and responses of point masses in the last few chapters and have found that the gravitational field couples to the energy-momentum tensor, which for a particle integrates to

$$T_{\mu\nu} = E \frac{dx_\mu}{dt} \frac{dx_\nu}{dt} .$$

In both the scalar and tensor cases the force is a function of velocity: in particular, a particle moving at right angles to a field $-\underline{\nabla}\varphi$ experiences a force

$$-m\,(1+v^2)\,\underline{\nabla}\varphi \qquad \text{tensor case}$$

$$-m\,(1-v^2)\,\underline{\nabla}\varphi \qquad \text{scalar case .}$$

We have tacitly been supposing that a composite bound system like an atom behaves in the same way as a particle: indeed it is clear that this must be the case from the Eötvös-Dicke experiments. In particular we saw in Chapter 2 that the gravitational and inertial masses associated with the kinetic energy of the nucleons in a nucleus must be equal to within 1 part in 10^{10}. The typical momentum of nucleons in a nucleus is given by $pr \sim \hbar$, with $r \sim 10^{-13}$ cm. The velocity of such a nucleon is $\sim 10^9 - 10^{10}$ cm s^{-1}. The relativistic correction to the gravitational force, acting on such a nucleon, clearly does not manifest itself when the nucleon is bound: something cancels these terms.

We have a hint as to what this could be. Considering only free particles, we postulated that the gravitational field couples to the conserved energy-momentum tensor. Now the energy-momentum tensor of a little piece of a bound system (like a nucleon in a nucleus) is not conserved. Rather than write

$$\frac{\partial}{\partial x_\mu} \mathcal{T}_{\mu\nu} = 0$$

we must write

$$\frac{\partial}{\partial x_\mu} \mathcal{T}_{\mu\nu} = \mathcal{F}_\nu \tag{6.3.1}$$

where $\mathcal{F}_\nu$ is the force density. The conserved tensor will be some construct $\mathcal{K}_{\mu\nu}$ such that

$$\frac{\partial}{\partial x_\mu} \mathcal{K}_{\mu\nu} = \frac{\partial}{\partial x_\mu} \mathcal{T}_{\mu\nu} - \mathcal{F}_\nu = 0 \tag{6.3.2}$$

and if gravitation is to couple to a conserved quantity, it is to $\mathcal{K}_{\mu\nu}$ that it must couple.

For a nucleon in a nucleus, the kinetic energy is not conserved, but the sum of kinetic and potential energy is conserved: it is to this that gravitation couples. What is the analogy for momentum?

Consider the nucleons in a nucleus as a gas confined within a box. For a single particle we have Eq. (5.4.4)

$$\frac{dp_\rho}{dt} = m\tilde{v}_\mu \tilde{v}_\nu \frac{\partial}{\partial x_\rho} h_{\mu\nu} - m\tilde{v}_\mu \tilde{v}_\nu \frac{\partial}{\partial x_\mu} h_{\rho\nu} .$$

For $d\underline{p}/dt$ we extract

$$\frac{d\underline{p}}{dt} = - m(1 + v^2)\, \underline{\nabla} \varphi + m\underline{v}\, \underline{v} \cdot \underline{\nabla} \varphi$$

where the term in $-m\underline{\nabla}\varphi$ comes from $T_{44}\, h_{44}$ and the other terms come from h_{ii} $(i = 1,2,3)$. The sum of $mv_i v_i$ over all nucleons is clearly not equal to zero, but reflects the mean square velocity. However, the sum over all nucleons of terms like $mv_i v_j$ denotes the rate at which momentum is incident in a given direction. At the walls of the nuclear box this momentum is turned around by the presence of stress in the walls (external forces as far as nucleons confined in a box are concerned). On integrating over the whole nucleus the sum of the pressure exerted by the nucleons and the stresses confining them must be zero for a bound system. For a system at rest the particle part of $\mathcal{K}_{ii}$ is just the pressure: gravity couples to both this and the stresses, and on integrating to get $\mathcal{K}_{ii}$, the two parts cancel for a bound system.

We can put the whole thing in a picturesque form by considering the equation

$$\frac{dp_\varphi}{dt} = - m(1 + a v^2)\, \underline{\nabla}\varphi + \underline{F} \tag{6.3.3}$$

where $a = +1$ for the tensor theory, and -1 for the scalar theory for a particle in a gravitational field. The force $\underline{F}$ represents the 'spring balance' from which it is hung: in equilibrium

$$\underline{F} = m(1 + a v^2)\, \underline{\nabla}\varphi .$$

Confine a gas of such particles in a box of volume V. Let there be n per unit volume. Then

$$\underline{F} = n m V(1 + a \overline{v^2})\, \underline{\nabla}\varphi \tag{6.3.4}$$

if we take no account of stress. The term $n m \overline{v^2}$ gives the pressure exerted on the walls $n m \overline{v^2} = 3P$.

If the box is lowered an infinitesimal distance into the potential its size changes (from the point of view of the Olympian reference frame) and consequently

$$V \to V(1 + 3a\,\Delta\varphi) \qquad \text{or} \qquad \Delta V = 3 V a \Delta\varphi$$

since $L \to L(1 + a\varphi)$ in a field φ. Then

$$P\,\Delta V = n m \overline{v^2}\, a V \Delta\varphi$$

$$= n m \overline{v^2}\, a V \underline{\nabla}\varphi \,.\, \Delta\underline{x} \,.$$

Thus

$$\underline{F} \,.\, \Delta\underline{x} = \left\{ n m V \underline{\nabla}\varphi \,.\, \Delta\underline{x} + P\,\Delta V \right\} \qquad (6.3.5)$$

represents the work done in moving an infinitesimal distance $\underline{x}$. It is only the first term on the right hand side that corresponds to weight, for the second term is cancelled by the stresses that hold the complex system together.

By making measurements in a laboratory within which the gravitational potential does not vary significantly, we cannot distinguish between the scalar and tensor theories, or indeed any combination of them. It is only by making measurements of relativistic corrections, over a region of space in which the gravitational potential changes substantially, that we can pick out the correct generalisation of Newton's theory of gravitation.

6.4 Covariant equations of motion

In a spherically symmetric gravitational field we have inferred that the equation of motion of a particle is

$$\frac{d\underline{v}}{dt} = -\,(1 + v^2)\, \underline{\nabla}\varphi + 4\underline{v}\,\underline{v}\,.\,\underline{\nabla}\varphi \,.$$

We would like to find the covariant equations corresponding to this. A crucial ingredient was our recognition of the slowing down of light which, again in a spherically symmetric field, we were able to express by setting $c \to \frac{c}{n}$ where $n = 1 - 2\varphi$.

In a more general case, it is best to start from the manifestly covariant equations

$$\frac{dp_\rho}{d\tau} = m_o \frac{dx_\mu}{d\tau}\frac{dx_\nu}{d\tau}\left\{ \frac{\partial}{\partial x_\rho} h_{\mu\nu} - \frac{\partial}{\partial x_\mu} h_{\rho\nu} \right\} \qquad (5.4.7a)\,(6.4.1)$$

which we derived in section 5.4 as a result of setting

$$E^2 - p^2 = -\,m_o^2 \qquad \text{and} \qquad \underline{p} = E\underline{v}$$

together with $E\frac{d\tau}{dt} = m_o$ (relations we have already been forced to modify).

In relativistic mechanics the momentum-energy four-vector can be written

$$p_\rho = m_o\frac{dx_\rho}{d\tau} \qquad d\tau^2 = -dx_\mu\, dx_\mu$$
$$p_i = m_o\frac{dx_i}{d\tau} \qquad E = m_o\frac{dt}{d\tau}\ .$$

With these identifications we may write

$$p_\rho p_\rho = m_o^2\frac{dx_\rho}{d\tau}\frac{dx_\rho}{d\tau} = -m_o^2$$

and

$$p_i = E\frac{d\tau}{dt}\frac{dx_i}{d\tau} = E\frac{dx_i}{dt} = E\, v_i\ ,$$

the standard results. Maintain these identifications and write

$$\frac{d^2x_\rho}{d\tau^2} = \frac{dx_\mu}{d\tau}\frac{dx_\nu}{d\tau}\left\{\frac{\partial h_{\mu\nu}}{\partial x_\rho} - \frac{\partial h_{\rho\nu}}{\partial x_\mu}\right\}\ . \tag{6.4.2}$$

This is a set of four equations from which we can extract the acceleration:

$$\frac{d^2x_i}{d\tau^2} = \frac{dx_\mu}{d\tau}\frac{dx_\nu}{d\tau}\left\{\frac{\partial h_{\mu\nu}}{\partial x_i} - \frac{\partial h_{i\nu}}{\partial x_\mu}\right\}$$

$$\frac{d^2x_4}{d\tau^2} = \frac{dx_\mu}{d\tau}\frac{dx_\nu}{d\tau}\left\{\frac{\partial h_{\mu\nu}}{\partial x_4} - \frac{\partial h_{4\nu}}{\partial x_\mu}\right\}\ .$$

Now

$$\frac{d^2x_i}{d\tau^2} = \frac{d}{d\tau}\left\{\frac{dx_i}{dt}\frac{dt}{d\tau}\right\} = \frac{d^2t}{d\tau^2}\frac{dx_i}{dt} + \frac{d^2x_i}{dt^2}\left(\frac{dt}{d\tau}\right)^2\ .$$

With $x_4 = it$ we obtain

$$\frac{d^2x_i}{dt^2}\left(\frac{dt}{d\tau}\right)^2 = \frac{dx_\mu}{d\tau}\frac{dx_\nu}{d\tau}\left\{\frac{\partial h_{\mu\nu}}{\partial x_i} - \frac{\partial h_{i\nu}}{\partial x_\mu}\right\}$$
$$+\, i\,\frac{dx_i}{dt}\frac{dx_\mu}{d\tau}\frac{dx_\nu}{d\tau}\left\{\frac{\partial h_{\mu\nu}}{\partial x_4} - \frac{\partial h_{4\nu}}{\partial x_\mu}\right\}$$

which can be written at once as

$$\frac{d^2x_i}{dt^2} = \frac{dx_\mu}{dt}\frac{dx_\nu}{dt}\left\{\frac{\partial h_{\mu\nu}}{\partial x_i} - \frac{\partial h_{i\nu}}{\partial x_\mu} + i\left[\frac{\partial h_{\mu\nu}}{\partial x_4} - \frac{\partial h_{4\nu}}{\partial x_\mu}\right]\frac{dx_i}{dt}\right\}\ . \tag{6.4.3}$$

Specialise to the time independent spherically symmetric field and find

$$\frac{d\underline{v}}{dt} = -(1+v^2)\,\underline{\nabla}\varphi + \underline{v}\,\underline{v}\cdot\underline{\nabla}\varphi - i\,\underline{v}\,\frac{dx_\mu}{dt}\frac{dx_\nu}{dt}\frac{\partial h_{4\nu}}{\partial x_\mu}$$
$$= -(1+v^2)\,\underline{\nabla}\varphi + 2\underline{v}\,\underline{v}\cdot\underline{\nabla}\varphi\ , \tag{6.4.4}$$

which is of course the answer we obtained setting $\underline{p} = E\underline{v}$, Eq. (5.4.9), which we found to be unsatisfactory. We must modify the kinematic equations in a covariant way.

Let us keep the relation

$$p_\rho p_\rho = -m_o^2$$

and set

$$p_\rho = m_o \chi_{\rho\sigma} \frac{dx_\sigma}{d\tau} .$$

We see from Eqs. (6.4.2) and (6.4.4) that we need to obtain equations which look something like

$$\frac{d^2x_\rho}{d\tau^2} = \frac{dx_\mu}{d\tau}\frac{dx_\nu}{d\tau}\left\{\frac{\partial h_{\mu\nu}}{\partial x_\rho} - 2\frac{\partial h_{\rho\nu}}{\partial x_\mu}\right\}$$

which is equivalent to

$$\frac{d^2x_\rho}{d\tau^2} = \frac{dx_\mu}{d\tau}\frac{dx_\nu}{d\tau}\left\{\frac{\partial h_{\mu\nu}}{\partial x_\rho} - \frac{\partial h_{\rho\nu}}{\partial x_\mu} - \frac{\partial h_{\mu\rho}}{\partial x_\nu}\right\}.$$

Then on differentiating the momentum with respect to τ, we have

$$\frac{dp_\rho}{d\tau} = m_o \chi_{\rho\sigma} \frac{d^2x_\sigma}{d\tau^2} + \frac{dx_\sigma}{d\tau} m_o \frac{d\chi_{\rho\sigma}}{d\tau}$$

$$= m_o \chi_{\rho\sigma} \frac{d^2x_\sigma}{d\tau^2} + m_o \frac{dx_\sigma}{d\tau}\frac{dx_\mu}{d\tau}\frac{d\chi_{\rho\sigma}}{dx_\mu} . \qquad (6.4.5)$$

Identify $\chi_{\rho\sigma} = \delta_{\rho\sigma} + h_{\rho\sigma}$ and obtain

$$(\delta_{\rho\sigma} + h_{\rho\sigma})\frac{d^2x_\sigma}{d\tau^2} = \frac{dx_\mu}{d\tau}\frac{dx_\nu}{d\tau}\left\{\frac{\partial h_{\mu\nu}}{\partial x_\rho} - 2\frac{\partial h_{\rho\nu}}{\partial x_\mu}\right\}. \qquad (6.4.6)$$

This is a manifestly covariant equation, and differs from our desired form only in the coefficient of the quantity $d^2x_\sigma / d\tau^2$. To first order in $h_{\mu\nu}$ we clearly obtain the equation of motion we are looking for. The correction term on the left hand side corresponds to an acceleration dependent four-force, determined by the gravitational potential.

We are now happy that the equation of motion (5.4.14) in a spherically symmetric potential characterised by φ is, to first order in φ, obtainable from a respectable general equation. Let us investigate the implications for the energy and momentum. We have

$$p_\rho p_\rho = -m_o^2$$

$$p_\rho = m_o \chi_{\rho\sigma} \frac{dx_\sigma}{d\tau} .$$

These are manifestly covariant equations: we may write

$$(\chi_{\rho\sigma})^{-1}\ p_\rho = m_o \frac{dx_\sigma}{d\tau}\ .$$

In the relatively simple case of diagonal $h_{\mu\nu}$ we have

$$p_i = m_o(1+h_{ii})\ \frac{dx_i}{d\tau}$$

$$E = m_o(1+h_{44})\ \frac{dt}{d\tau}\ .$$

Then

$$\underline{p} = (1+h_{ii})(1+h_{44})^{-1}E\frac{d\underline{x}}{dt}$$

and if we have $p^2 - E^2 = -m_o^2$, then

$$E = \frac{m_o}{\sqrt{1-(1+h_{ii})^2(1+h_{44})^{-2}\ v^2}}$$

which again agrees with our previous identifications. A related, but much more complicated equation would obtain for general $h_{\mu\nu}$.

It is worth noting the implication of the relations

$$p_\rho p_\rho = -\ m_o^2\ \ ;\qquad p_\rho = m_o\ \chi_{\rho\sigma}\ \frac{dx_\sigma}{d\tau}\ .$$

These can only be true, with m_o an Olympian frame invariant, if

$$\chi_{\rho\sigma}dx_\sigma\chi_{\rho\alpha}\ dx_\alpha = -\ d\tau^2\ . \qquad (6.4.7)$$

We recover the standard kinematic relations of special relativity on effecting a redefinition of coordinates

$$\chi_{\rho\sigma}\ dx_\sigma \rightarrow dx_\rho$$

which matches with our earlier identifications $L_\varphi \rightarrow L(1+\varphi)$ and $T_\varphi \rightarrow T(1-\varphi)$ for the time independent spherically symmetric field: a length L_φ is measured in the Olympian frame and is $L_\varphi = (1+\varphi)L$ where L is measured very far away from the source of φ. The quantity

$$dx_i = \chi_{i\sigma}\ dx_\sigma$$

is $\chi_{i\sigma}L_\varphi = (1-\varphi)L_\varphi = L$ (to first order in φ). Write Eq. (6.4.7) in the form

$$(1+h_{ii})^2\ d\underline{x}^2 - (1+h_{44})^2\,dt^2 = -\ d\tau^2\ .$$

If $(d\tau)^2 = 0$ as it is for light in standard special relativity, then

$$\left|\frac{dx}{dt}\right| = (1 + h_{44})(1 + h_{ii})^{-1}$$
$$\simeq (1 + 2\varphi)$$

for the tensor theory. More generally

$$(1 + h_{ii})^2 \left(\frac{dx}{dt}\right)^2 - (1 + h_{44})^2 = - \left(\frac{d\tau}{dt}\right)^2 .$$

So that

$$\Delta t = \frac{\Delta\tau}{\sqrt{(1 + h_{44})^2 - (1 + h_{ii})^2 v^2}}$$
$$= \frac{(1 + h_{44})^{-1} \Delta\tau}{\sqrt{1 - (1 + h_{ii})^2 (1 + h_{44})^{-2} v^2}} \tag{6.4.8}$$

which is the usual time dilation result modified by the factor $(1 + h_{44})^{-1}$.

We may also note that with $d\tau^2 = 0$

$$(1 + h_{11})^2 dx^2 + (1 + h_{22})^2 dy^2 + (1 + h_{33})^2 dz^2 = (1 + h_{44})^2 dt^2$$

for the case of a diagonal $h_{\mu\nu}$ but not necessarily a spherically symmetric field. The velocity of light in the x direction is not necessarily equal to the velocity of light in the y direction: it is clear that a non-zero h_{11} affects lengths in x and so on. Indeed we can see that the scalar, Newtonian and tensor theories slow clocks down in the same way $(h_{44} = \varphi)$ while the Newtonian theory leaves lengths unchanged $(h_{ii} = 0)$. The tensor field shrinks lengths $(h_{ii} = -\varphi)$ while the scalar field expands them $(h_{ii} = \varphi)$.

We may now proceed one step further. If we write

$$p_{\varphi_\rho} = (\delta_{\rho\sigma} + 2h_{\rho\sigma})\, m_o \frac{dx_\sigma}{d\tau} \tag{6.4.9}$$

and

$$\frac{dp_{\varphi_\rho}}{d\tau} = m_o \frac{dx_\mu}{d\tau} \frac{dx_\nu}{d\tau} \frac{dh_{\mu\nu}}{\partial x_\rho} \tag{6.4.10}$$

we obtain for the acceleration equations in free-fall

$$(\delta_{\rho\sigma} + 2h_{\rho\sigma}) \frac{d^2x_\sigma}{d\tau^2} = \frac{dx_\mu}{d\tau} \frac{dx_\nu}{d\tau} \left\{ \frac{\partial h_{\mu\nu}}{\partial x_\rho} - \frac{\partial h_{\rho\nu}}{\partial x_\mu} - \frac{\partial h_{\mu\rho}}{\partial x_\nu} \right\} \tag{6.4.11}$$

which to first order in $h_{\mu\nu}$ give us once more the equation of motion in a spherically symmetric field. The advantage of this identification is that in the absence of external forces other than gravitational the quantity

$$p_{\varphi_4} = (\delta_{4\sigma} + 2h_{4\sigma})\, m_o \frac{dx_\sigma}{d\tau} \tag{6.4.12}$$

is conserved so long as $\partial h_{\mu\nu}/\partial x_4$ vanishes and

$$p_{\varphi_i} = (\delta_{i\sigma} + 2h_{i\sigma})\, m_0 \frac{dx_\sigma}{d\tau} \qquad (6.4.13)$$

is conserved so long as $\partial h_{\mu\nu}/\partial x_i$ vanishes. It is these quantites that behave like momentum and energy when external forces (for example electromagnetic forces) are applied in the presence of a gravitational potential. For the case of diagonal $h_{\mu\nu}$

$$E_\varphi = \frac{m_0(1 + h_{44})}{\sqrt{1 - (1 + h_{ii})^2\ (1 + h_{44})^{-2}\ v^2}} \qquad (6.4.14)$$

$$p_{\varphi_i} = \frac{(1 + h_{ii})^2\ (1 + h_{44})^{-1}\ m_0\ v_i}{\sqrt{1 - (1 + h_{ii})^2\ (1 + h_{44})^{-2}\ v^2}}\ . \qquad (6.4.15)$$

Our calculations in Chapter 5 and Chapter 6.1 and 6.2, augmented by the considerations of this section, demonstrate the machinery by which the gravitational field inserts the extra factors not contained in special relativity. The changes of scale implied by (6.4.7) must not be added to those calculated with modified expressions for mass and energy, for they are a different statement of the same physical effects.

6.5 The Lagrangian formalism: again for experts

The manifestly covariant equation (6.4.11) embodies all the results of Chapter 5 and Chapter 6, sections 1 and 2 in a more general form. This equation may be derived by writing the invariant Lagrangian for a particle in an external gravitational field as

$$L = \tfrac{1}{2}\, m_0 \frac{dx_\mu}{d\tau}\frac{dx_\mu}{d\tau} + m_0 \frac{dx_\mu}{d\tau}\frac{dx_\nu}{d\tau} h_{\mu\nu} \qquad (6.5.1)$$

and feeding it into the covariant Euler-Lagrange equation (3.3.3): the acceleration dependent forces emerge quickly and naturally in this formulation, which is based on the identification of the interaction

$$h_{\mu\nu}\ \mathfrak{T}_{\mu\nu}$$

with an invariant Lagrangian density.

The modifications to the equations of electrodynamics necessitated by the presence of gravitation can also be worked out by augmenting the Lagrangian function for the free fields plus sources [1] with an interaction term, the gravitational potential $h_{\mu\nu}$ being coupled to the energy-momentum tensor of the electromagnetic field [2,3]. This approach both generalises our results to the general tensor field and automatically provides covariant equations of electrodynamics in the presence of gravitation. The results are valid for weak gravitational fields but can be extended to strong fields [2], yielding in the end Einstein's theory.

References

[1] See for example L. Landau and E.M. Lifshitz, 'The Classical Theory of Fields' (Pergamon, 1962) Chapter 4.

[2] W.E. Thirring, Annals of Physics, 16, 96 (1961). This is also a general reference for this chapter.

[3] In the context of conventional general relativity, see for example:

S. Weinberg, 'Gravitation and Cosmology', (Wiley 1972) Chapters 5 and 12.

L. Landau and E.M. Lifschitz, 'The Classical Theory of Fields' (Pergamon 1962) Section 88.

CHAPTER 7

THE PRECESSION OF THE PERIHELION OF MERCURY

7.1 Introduction

In the previous chapter we have developed a picture of gravity which satisfies the uniqueness of acceleration in a gravitational field and yields the gravitational redshift, deflection of light by the Sun and radar echo delay in accord with the predictions of the general theory of relativity and experiment. We have also seen how such a theory can contain a principle of equivalence of physics in all local freely falling frames. We have thus dealt with three of the four famous tests of general relativity (or indeed any theory of gravitation). The fourth of these tests is the precession of the perihelion of Mercury. This phenomenon is of particular importance because it is the only currently available test of the nonlinear terms in this formulation of the theory of gravitation.

A planet in orbit in a perfect inverse square field follows, in Newtonian mechanics, a path that is an ellipse, with axes fixed with respect to the absolute space of Newtonian mechanics, operationally defined as the reference frame provided by the very distant stars. If the gravitational field in which the planet moves is not perfect inverse square, this is no longer true and, among other effects, the axes of the ellipse slowly rotate with respect to the distant stars. The perihelion of an orbit is the point of closest approach to the Sun and if the ellipse is rotating, the perihelion is slowly rotating with respect to the distant stars. (In the solar system the principal cause of such rotation is planetary perturbation.) The easiest way of seeing what is going on is to consider circular motion about the Sun (or

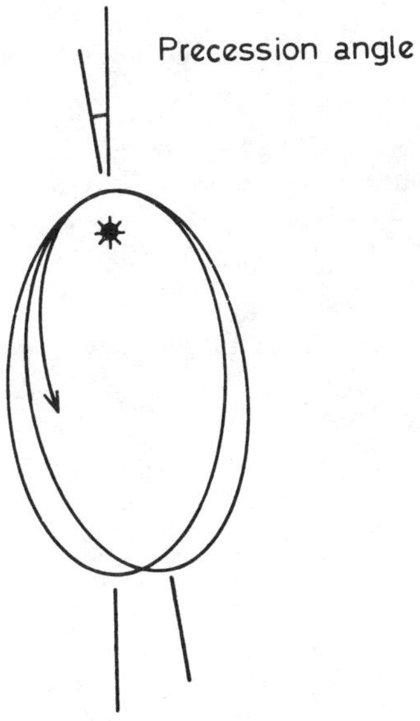

Fig.7.1.1 The trajectory of a test particle in a precessing orbit (of extreme eccentricity) is shown for $\sim 2\frac{1}{4}$ revolutions. The precession is greatly exaggerated

any other primary) and superimpose on this circular motion a small radial oscillation. If the periods for these two superimposed motions coincide, the planet returns to perihelion after sweeping out 2π radians. In this very special case the orbit is closed and repeats every revolution. If the periods are not the same, but are commensurate, the orbit repeats after a certain number of revolutions, while if the two periods are not commensurate, the orbit never repeats.

If the period of the radial vibrations is T_R and the period for a rotation of 2π is T_θ, then the planet sweeps out $2\pi\,(T_R)/T_\theta$ radians in time T_R. The angular advance of the perihelion each revolution is thus

$$2\pi\left\{\frac{T_R}{T_\theta} - 1\right\} \text{ radians}$$

(see Fig.7.1.1).

7.2 Perihelion advance in Newtonian mechanics

We will first demonstrate that in Newtonian mechanics there is no precession of the perihelion for motion in a pure $1/r$ potential.

The gravitation potential is

$$\varphi = -\frac{GM}{r}$$

and the equation of motion is

$$\frac{d\underline{v}}{dt} = -\underline{\nabla}\varphi = -\frac{GM}{r^2}\frac{\underline{r}}{r} . \tag{7.2.1}$$

The orbit is planar and resolving (7.2.1) into radial and tangential components we have

$$\frac{d^2r}{dt^2} - r\left(\frac{d\theta}{dt}\right)^2 = -\frac{GM}{r^2} \tag{7.2.2}$$

$$r\frac{d^2\theta}{dt^2} + 2\frac{dr}{dt}\frac{d\theta}{dt} = 0 \tag{7.2.3}$$

or

$$\frac{1}{r}\frac{d}{dt}\left(r^2\frac{d\theta}{dt}\right) = 0 . \tag{7.2.4}$$

We can write Eq. (7.2.4) as

$$r^2\frac{d\theta}{dt} = h \tag{7.2.5}$$

where h is a constant and is clearly the angular momentum per unit mass, since the angular velocity is $d\theta/dt$. Eqs.(7.2.3), (7.2.4) thus correspond to conservation of angular momentum.

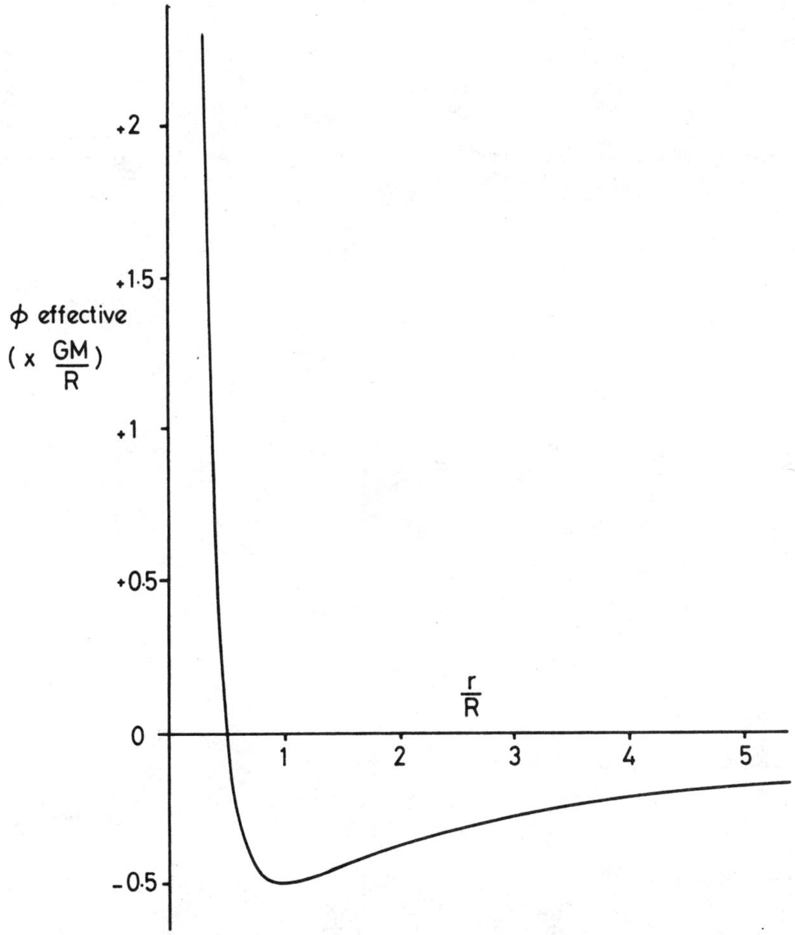

Fig.7.2.1 The effective potential governing the radial component of motion in a $1/r$ potential. A circular orbit corresponds to $r = R$. Small oscillations about this point represent a nearly circular orbit

Since h is a constant of the motion we can decouple the angular and radial equations by writing (7.2.2) in the form

$$\frac{d^2r}{dt^2} = \frac{h^2}{r^3} - \frac{GM}{r^2} \tag{7.2.6}$$

corresponding to one dimensional motion in the potential

$$\varphi_{effective} = \frac{h^2}{2r^2} - \frac{GM}{r} \tag{7.2.7}$$

illustrated in Fig.7.2.1. The first term in Eq. (7.2.7) is a centrifugal repulsion of exactly the kind encountered in quantum mechanics problems.

For circular motion
$$\left.\frac{d^2r}{dt^2}\right|_R = 0$$

and so

$$\frac{h^2}{R^3} = \frac{GM}{R^2} \tag{7.2.8}$$

where R is the radius of the circular orbit. Introduce a radial oscillation by setting

$$r = R + \rho$$

when Eq. (7.2.6) becomes

$$\frac{d^2\rho}{dt^2} = \frac{h^2}{(R+\rho)^3} - \frac{GM}{(R+\rho)^2} .$$

Expanding in powers of ρ/R

$$\frac{d^2\rho}{dt^2} = \left\{- \frac{3h^2}{R^4} + \frac{2GM}{R^3}\right\} \rho$$

and substituting for h^2 from Eq. (7.2.8)

$$\frac{d^2\rho}{dt^2} = - \frac{GM}{R^3} \rho . \tag{7.2.9}$$

Then

$$\rho = \rho_o \sin \sqrt{\frac{GM}{R^3}}\, t$$

and the period of small radial oscillations is

$$T_R = 2\pi \sqrt{\frac{R^3}{GM}} . \tag{7.2.10}$$

The tangential velocity is

$$r \frac{d\theta}{dt} = \frac{h}{r} \simeq \sqrt{\frac{GM}{r}} \simeq \sqrt{\frac{GM}{R}}$$

and the distance round the orbit is approximately $2\pi R$, so the rotational period is

$$T_\theta = 2\pi \sqrt{\frac{R^3}{GM}} \tag{7.2.11}$$

which is identical with the radial period. There is no advance of the perihelion.

It should be clear that this is a special case. Suppose that the $1/r$ potential is perturbed by a (central) potential with different radial dependence :

$$\varphi = - \frac{GM}{r} \left\{1 + \frac{a}{r^n}\right\} .$$

Then for circular motion

$$\frac{h^2}{R^3} = \frac{GM}{R^2} \left\{1 + \frac{(n+1)a}{R^n}\right\}$$

and the radial equation is

$$\frac{d^2\rho}{dt^2} = \frac{h^2}{(R+\rho)^3} - \frac{GM}{(R+\rho)^2}\left\{1 + \frac{(n+1)a}{(R+\rho)^n}\right\}$$

$$\simeq -\frac{GM}{R^3}\left\{1 - \frac{(n^2-1)a}{R^n}\right\}\rho$$

and

$$T_R = 2\pi\sqrt{\frac{R^3}{GM}}\left\{1 + \tfrac{1}{2}\frac{(n^2-1)a}{R^n}\right\}$$

$$T_\theta = 2\pi\sqrt{\frac{R^3}{GM}}\left\{1 - \tfrac{1}{2}\frac{(n+1)a}{R^n}\right\}$$

(with $n=0$ the two periods are the same, of course). The advance of the perihelion is

$$2\pi\left\{\frac{T_R}{T_\theta} - 1\right\} = \pi n(n+1)\frac{a}{R^n}. \tag{7.2.12}$$

The only example of this kind of simple perturbation that is of interest is for $n=2$, corresponding to a quadrupole field in the equatorial plane.

7.3 The relativistic theory and the need for nonlinear terms

In Chapter 5 we found an equation of motion

$$\frac{d\underline{v}}{dt} = -(1+v^2)\underline{\nabla}\varphi + 4\underline{v}\,\underline{v}\cdot\underline{\nabla}\varphi \qquad (5.4.14) \qquad (7.3.1)$$

which reduces to (7.2.1) as $v \to 0$ and gives the correct deflection of light by the Sun. This equation was derived working only to first order in the potential. As it stands it clearly contains departures from Newtonian mechanics of order v^2 (with $c=1$) and we could at once apply the methods of the previous paragraph to find the precession rate of a planetary orbit due to these terms.

However, for planetary motion

$$\frac{v^2}{r} \sim \frac{GM}{r^2}$$

and so the dimensionless quantities v^2/c^2 and GM/rc^2 are of the same order of magnitude. If we are to work out the effects of relativistic gravitation on planetary orbits, we must include corrections $\sim \varphi\underline{\nabla}\varphi$ along with those $\sim v^2\,\underline{\nabla}\varphi$. We have already seen that the equations of the gravitational field are expected to be nonlinear, giving rise to terms $\sim \varphi\underline{\nabla}\varphi$ in any equation of motion. We shall therefore study the advance of the perihelion obtaining with an equation of motion:

$$\frac{d\underline{v}}{dt} = -(1+v^2)\underline{\nabla}\varphi - \alpha\underline{\nabla}\varphi^2 + 4\underline{v}\,\underline{v}\cdot\underline{\nabla}\varphi \tag{7.3.2}$$

where φ has been replaced by $\varphi(1+\alpha\varphi)$ in (7.3.1) and terms in $v^2\varphi^2$, being $0(\varphi^3)$, have been dropped. Our first task is to discover the appropriate value of the coefficient α. Some of this coefficient comes from the nonlinearity of the field equations, but we should note that even neglecting this nonlinearity we find $\alpha=1$ from the most respectable looking formulation of the free-fall equations in the tensor theory, Eq. (6.4.11). (This result is easily obtained on applying to (6.4.11) the procedures used to extract (6.4.4) from (6.4.2) and keeping all terms up to second order in small quantities.)

7.4 Strong equivalence and the nonlinear terms

We are not going to attempt to develop a consistent nonlinear theory to even second order in the potential φ. We shall cut this Gordian knot, instead of unravelling it, by applying a form of the principle of strong equivalence.

We saw in Chapter 2 that inertial and gravitational mass are strictly proportional to very high accuracy for energy contained in electromagnetic and strong interaction fields. The Eötvös-Dicke experiments have little to say about the weak interactions and nothing to say about the inertial and gravitational masses of gravitational fields themselves. A principle of equivalence states that in any local freely falling frame the same Lorentz covariant laws of physics apply, with the same numerical content. Such principles represent an abstraction from special relativity plus the results of the Eötvös-Dicke experiments and so are reasonably sound for the laws of electromagnetism and strong interaction physics but represent an assumption for the laws of gravity. The principle of strong equivalence assumes that all laws of physics, including gravity itself, are the same in any local freely falling frame. It was this postulate that Einstein used to arrive at the field equations of general relativity: we shall be less ambitious and use it to find the value of α in Eq.(7.3.2), assuming that a completely consistent tensor theory would contain the principle of strong equivalence (as is in fact the case).

If the principle of strong equivalence holds, then on gently lowering a small solar system into a gravitational potential on a much larger scale, say a galactic potential, the characteristic dimensions and frequencies of the system must remain the same, measured locally. In the Olympian reference system atoms shrink and vibrate more slowly, light is slowed down and so the small solar system must shrink and slow down in proportion. As we lower our little solar system into the galactic potential Φ we expect its linear dimensions to change as $1+\Phi$ and its characteristic times to change as $1-\Phi$. Velocities will thus vary as $1+2\Phi$, and the quantity

$$\frac{v^2}{r} \simeq 1 + 3\Phi \ .$$

Now

$$\frac{v^2}{r} \sim \frac{\varphi}{r}$$

where φ is the internal potential of the little solar system and so we must have

$$\varphi \sim 1 + 4\Phi$$

that is

$$\varphi \rightarrow \varphi(1+4\Phi) \ .$$

The same result may be obtained by noting that the energy of a body of mass m in the potential φ is $-m\varphi$ and that m varies as $1-3\Phi$, while energy must vary as $1+\Phi$ for the principle of strong equivalence to hold.

Now the quantity $2\varphi\Phi$ is the cross product term in $(\varphi+\Phi)^2$ so consideration of the properties of a small gravitational system in a large scale potential suggests the general replacement

$$\Psi \rightarrow \Psi + 2\Psi^2$$

as the second order correction to a potential Ψ in Eq. (7.3.1).

This makes sense. Consider two masses m in a field Ψ, separated by a potential difference $\Delta\Psi$. Their energy difference is $m\Delta\Psi$. Compare this with the difference between atomic energy levels at Ψ. This latter difference changes with Ψ as $1+\Psi$ and so the principle of strong equivalence is satisfied if $m\Delta\Psi$ changes as $1+\Psi$. Since m varies as $1-3\Psi$, $\Delta\Psi$ varies as $1+4\Psi$. That is,

$$\Delta\Psi \rightarrow \Delta\Psi(1+4\Psi) = \Delta(\Psi+2\Psi^2) \ .$$

In order to compute planetary motion to second order in small quantities, under the assumption of strong equivalence, we must replace the potential $\varphi(=-\frac{GM}{r})$ by the quantity $\varphi+2\varphi^2$ in equation (7.3.1); that is, set $\alpha=2$ in Eq.(7.3.2).

7.5 Calculation of the advance of perihelion

We take as our equation of motion

$$\frac{d\underline{v}}{dt} = -(1+v^2)\,\underline{\nabla}\varphi - \alpha\underline{\nabla}\varphi^2 + 4\underline{v}\;\underline{v}\cdot\underline{\nabla}\varphi \qquad (7.3.2) \qquad (7.5.1)$$

and begin by resolving into components.

The angular equation is

$$\frac{1}{r}\frac{d}{dt}\left(r^2\frac{d\theta}{dt}\right) = 4r\left(\frac{d\theta}{dt}\right)\frac{dr}{dt}\frac{\partial\varphi}{\partial r} \qquad (7.5.2)$$

It is important to notice that this equation, which corresponds to conservation of angular momentum, differs from the Newtonian equivalent, in that the right hand side is non-zero.

Since the potential φ is not explicitly time dependent, we may write (7.5.2) as

$$\frac{1}{r}\frac{d}{dt}\left(r^2\frac{d\theta}{dt}\right) = 4r\left(\frac{d\theta}{dt}\right)\frac{d\varphi}{dt} . \tag{7.5.3}$$

Set

$$x r^2\frac{d\theta}{dt} = h \tag{7.5.4}$$

where h is a constant and obtain an equation for the relativistic correction factor x. Differentiate (7.5.4) with respect to time:

$$x\frac{d}{dt}\left(r^2\frac{d\theta}{dt}\right) + r^2\frac{d\theta}{dt}\frac{dx}{dt} = 0 . \tag{7.5.5}$$

Equations (7.5.3) and (7.5.5) yield

$$r^2\frac{d\theta}{dt}\frac{dx}{dt} + x\left\{4r^2\frac{d\theta}{dt}\frac{d\varphi}{dt}\right\} = 0$$

or

$$\frac{dx}{dt} = -4x\frac{d\varphi}{dt} .$$

Since we need to work only to one order in small quantities higher than the Newtonian limit, it is sufficiently accurate to set

$$x = 1-4\varphi$$

(x must reduce to 1 as $\varphi \to 0$) and the equation for angular motion becomes

$$r^2\frac{d\theta}{dt} \simeq h\left\{1-4\frac{GM}{r}\right\}. \tag{7.5.6}$$

Since we earlier made the identification

$$m = \frac{(1-3\varphi)\,m_o}{\sqrt{1-(1-2\varphi)^2 v^2}}$$

(see Eq. (5.5.5)) it might be thought that the quantity

$$(1-3\varphi)\,r^2\frac{d\theta}{dt}$$

would be conserved. This is not the case. The angular momentum $\underline{J}$ is

$$\underline{J} = m\underline{v}\times\underline{r} .$$

Since

$$\underline{v}\times\frac{d\underline{r}}{dt} = \underline{v}\times\underline{v} = 0 ,$$

$$\frac{d\underline{J}}{dt} = (\underline{v}\times\underline{r})\frac{dm}{dt} + m\left(\frac{d\underline{v}}{dt}\times\underline{r}\right) .$$

Equations (7.3.1) and (7.5.1) yield:

$$\frac{d\underline{v}}{dt}\times\underline{r} = 4\ (\underline{v}\times\underline{r})\ \underline{v}\cdot\underline{\nabla}\varphi$$

so

$$\frac{d\underline{J}}{dt} = (\underline{v}\times\underline{r})\left\{\frac{dm}{dt} + 4\,m\ \underline{v}\cdot\underline{\nabla}\varphi\right\}.$$

To first order in small quantities,

$$m = m_0\left\{1 - 3\varphi + \tfrac{1}{2}v^2\right\}$$

and so

$$\frac{dm}{dt} = m_0\left\{\underline{v}\cdot\frac{d\underline{v}}{dt} - 3\underline{v}\cdot\underline{\nabla}\varphi\right\}.$$

The leading term in the expressions for $d\underline{v}/dt$ is $-\underline{\nabla}\varphi$ so that to first order in φ, $d\underline{J}/dt$ is indeed zero, with $\underline{v}\times\underline{r}$ given by Eq. (7.5.6).

The radial equation of motion obtained from Eq. (7.5.1) is

$$\frac{d^2r}{dt^2} - r\left(\frac{d\theta}{dt}\right)^2 = -\left\{1 + \left(\frac{dr}{dt}\right)^2 + \left(r\,\frac{d\theta}{dt}\right)^2\right\} + \frac{2\alpha}{r}\left(\frac{GM}{r}\right)^2 + 4\left(\frac{dr}{dt}\right)^2\frac{GM}{r^2}\,. \tag{7.5.7}$$

For a circular orbit both dr/dt and d^2r/dt^2 are zero. For small radial oscillations superimposed on such a circular orbit, the terms in $(dr/dt)^2$ will be very small. They are also anharmonic, do not affect the basic period and will cancel out over a few revolutions. We therefore drop them and obtain

$$\frac{d^2r}{dt^2} - r\left(\frac{d\theta}{dt}\right)^2 \simeq -\left\{1 + \left(r\,\frac{d\theta}{dt}\right)^2\right\}\frac{GM}{r^2} + \frac{2\alpha}{r}\left(\frac{GM}{r}\right)^2\,. \tag{7.5.8}$$

We now use Eq. (7.5.6) to substitute for $d\theta/dt$. Remembering that

$$\frac{h^2}{R^2} \sim \frac{GM}{R}$$

and that we need only keep terms up to $0\left(\frac{GM}{R}\right)^2$, Eq. (7.5.8) becomes

$$\frac{d^2r}{dt^2} \simeq -\left\{1 + \frac{h^2}{r^2}\right\}\frac{GM}{r^2} + \frac{2\alpha}{r}\left(\frac{GM}{r}\right)^2 + \frac{h^2}{r^3}\left(1 - \frac{8GM}{r}\right)\,. \tag{7.5.9}$$

We now set $r = R + \rho$, with R given by

$$\left.\frac{d^2r}{dt^2}\right|_R = 0\ ,\quad \left.\frac{dr}{dt}\right|_R = 0$$

and obtain to order $\left(\frac{GM}{r}\right)^2$

$$\frac{d^2\rho}{dt^2} = -\frac{GM}{R^3}\left\{1 - \frac{9GM}{R}\right\}\rho$$

and hence a vibrational period

$$T_R = 2\pi\sqrt{\frac{R^3}{GM}}\left\{1 + \frac{9}{2}\frac{GM}{R}\right\}\,. \tag{7.5.10}$$

The terms in α disappear on substituting the value of h obtained from Eq. (7.5.9) with $d^2r/dt^2 = 0$, which is

$$\frac{h^2}{R^2} = \frac{GM}{R}\left\{1 + (9 - 2\alpha)\frac{GM}{R}\right\}. \tag{7.5.11}$$

The orbital velocity is given by Eq. (7.5.6)

$$R\frac{d\theta}{dt} = \frac{h}{R}\left\{1 - \frac{4GM}{R}\right\} = \sqrt{\frac{GM}{R}}\left\{1 + (\tfrac{1}{2} - \alpha)\frac{GM}{R}\right\}$$

and so the angular period is

$$T_\theta = 2\pi\sqrt{\frac{R^3}{GM}}\left\{1 - (\tfrac{1}{2} - \alpha)\frac{GM}{R}\right\}. \tag{7.5.12}$$

The advance of the perihelion each revolution is thus

$$2\pi\left\{\frac{T_R}{T_\theta} - 1\right\} = 2\pi(5 - \alpha)\frac{GM}{R} \tag{7.5.13}$$

radians.

With $\alpha = 2$, the value we inferred by applying the principle of strong equivalence, we obtain $6\pi(GM)/R$ radians per revolution, which is Einstein's result. With $\alpha = 1$, yielded by the tensor theory without any effects of self-interaction, we obtain $8\pi(GM)/R$ radians per revolution, 4/3 of the result from general relativity.

7.6 The precession of the perihelion of Mercury

The planet lying closest to the Sun is Mercury. The orbit of Mercury is thus most sensitive to the relativistic corrections we have been discussing, and in addition has the advantage of high eccentricity, $e = 0.206$. This eccentricity does not affect the rate of advance of perihelion, to the accuracy of Eq. (7.5.13), but it does make the measurement of this advance easier. It has been known for over a century that after the effects of planetary perturbations have been removed, the perihelion of Mercury retains an anomalous advance of 43" (seconds of arc) per century.

Our expression for the advance per revolution is, after inserting explicitly a factor c^2 which had previously been set equal to 1 for convenience

$$6\pi\frac{GM_\odot}{R_m c^2}$$

where R_m is the appropriate radius parameter for the orbit of Mercury.

The orbit of Mercury has a semi-major axis $a = 0.3871$ a.u. where 1 a.u. (astronomical unit) is 1.496×10^{13} cm. The eccentricity of the orbit is

$$e = 0.206$$

and so the semi-minor axis

$$b = a\sqrt{1-e^2}$$

is 2% shorter than the semi-major axis. Our simple treatment in terms of small radial vibrations does not tell us directly what quantity to take for R_m. However, we note that the advance of the perihelion is given by

$$-6\pi\frac{\varphi_R}{c^2},$$

where φ_R is the potential at R. The elliptical orbit means that a small range of potentials is sampled in the course of the orbit and so we shall replace φ_R by $\langle\varphi\rangle$, the average value of the potential, and hence evaluate

$$6\pi\frac{GM_\odot}{c^2}\langle\frac{1}{r}\rangle .$$

The equation of an ellipse is

$$\frac{\ell}{r} = 1 + e\cos\theta$$

where θ is measured from the major axis, so averaging over θ, $\langle\frac{1}{r}\rangle = \frac{1}{\ell}$ where ℓ is the parameter known as the semi-latus rectum (the distance across the orbit, normal to the major axis, measured through the focus) and is given by

$$\ell = a(1-e^2)$$

or equivalently by

$$\frac{1}{\ell} = \frac{1}{2}\left(\frac{1}{r_{max}} + \frac{1}{r_{min}}\right).$$

Our final expression for the advance of the perihelion is thus

$$6\pi\frac{GM_\odot}{\ell c^2} \tag{7.6.1}$$

which is the result obtained by a more complete treatment of Eq. (7.5.1) with $\alpha = 2$, which can itself be derived from general relativity, [1].

With

$$\ell = 0.555 \times 10^{13}\ \text{cm}$$
$$M_\odot = 1.99 \times 10^{33}\ \text{gm}$$

we obtain

$$6\pi\frac{GM_\odot}{\ell c^2} = 0.501 \times 10^{-6}$$

radians per revolution. The period of Mercury is 0.241 years and so the rate of advance of the perihelion of Mercury given by the expression (7.6.1) is 43" per century.

The precession of the perihelion of Mercury is measured with respect to an Earth based frame of reference. This frame cannot be directly related to the frame of the distant stars because the rotation axis of the Earth does not remain fixed in such a frame but precesses with a period of about 26,000 years (the precession of the equinoxes) due to tidal gravitational forces acting on the quadrupole moment of the Earth. The observed precession of the perihelion of Mercury (~ 5600" ± ~ 0.5" per century) is made up from a piece due to the precession of the equinoxes (~ 5025" per century), a piece due to planetary perturbations (~ 532" per century) and the residual anomalous precession which [2] is ~ 43" ± 1" which is in agreement with general relativity (otherwise known as the massless self-interacting spin 2 theory). The advance of the perihelion has also been measured for other planets but the results are accurate to no better than 10%.

The agreement between the calculated and observed values of the centennial advance of the perihelion of Mercury thus provides a verification at the 2% level of the equation of motion (7.5.1) with $\alpha = 2$, rather than $\alpha = 1$ as yielded by the spin 2 theory before allowing for additional nonlinear terms due to selfcoupling, and is the only test of these nonlinear terms *. This conclusion is however only warranted if nothing has been left out: in section 7.7 we discuss the quadrupole moment of the Sun.

The recent discovery of a pulsar in a binary system [3] offers the possibility of the investigation of much stronger gravitational fields than are easily accessible in the solar system. Pulsar PSR 1913 + 16 has a pulse period of 59 ms and is a member of a binary system with a period of 7.75 hours. The orbit is highly eccentric. The pulsar is presumably a neutron star with a mass $\sim M_\odot$ and the masses of the two components cannot be too different. There is evidence that the companion is much more condensed than a main sequence star and is therefore a white dwarf, a neutron star or possibly even a black hole.

The quantity

$$\frac{GM}{Rc^2} \approx 10^{-6}$$

thus making in principle the relativistic Doppler shift and gravitational redshift variations in pulse frequency easily observable, and giving a periastron advance of several degrees a year. There is also the possibility of observing the effects of spin-orbit coupling.

---

* It is however possible to construct a theory of gravity without self-interaction which yields the correct value for the advance of the perihelion of Mercury [S. Deser and B. Laurent, Ann. Phys., 50, 76 (1968)]

7.7 The oblateness of the Sun

In section 7.2 we noted that a precession of the perihelion would be induced, in the framework of Newtonian mechanics, by a perturbing potential varying with r as $r^{-(n+1)}$, with n non-zero. Such terms would be introduced if the Sun were not perfectly spherically symmetric, the only term of importance being the potential due to a quadrupole moment, varying with r as r^{-3}. A quadrupole moment would be reflected in the figure of the Sun: the disc of the Sun would be slightly oblate (flattened at the poles). Some oblateness is expected anyway because of the Sun's rotation. Interest in this possibility was reawakened some ten years ago as a result of measurements by Dicke and Goldenberg which implied a quadrupole moment many times greater than expected on the basis of a uniformly rotating Sun and sufficient to contribute some 3" per century to the advance of the perihelion of Mercury [4].

In Newtonian theory the gravitational potential of a body of local mass density ρ is

$$\varphi(\underline{r}) = - G \int \frac{\rho(\underline{r}')\, d^3r'}{|\underline{r} - \underline{r}'|} \quad . \tag{7.7.1}$$

Expanding as far as the quadrupole term yields for a rotationally symmetric body

$$\varphi(\underline{r}) \simeq - \frac{GM}{r} - \frac{G}{r^3}\left(\frac{3}{2}\cos^2\theta - \frac{1}{2}\right)\int \rho(\underline{r}')\left(\frac{3}{2}\cos^2\Theta - \frac{1}{2}\right) r'^2\, d^3r' \tag{7.7.2}$$

where θ is the polar angle of the vector $\underline{r}$ and Θ the polar angle of the vector $\underline{r}'$, both measured from the axis of symmetry of the quadrupole. It is convenient to write this as

$$\varphi(\underline{r}) = - \frac{GM}{r}\left\{1 - P_2(\cos\theta)\, Q\left(\frac{r_0}{r}\right)^2\right\} \tag{7.7.3}$$

where Q is a quadrupole moment parameter defined by

$$Q = - \frac{1}{Mr_0^2} \int \rho(\underline{r}')\left(\frac{3}{2}\cos^2\Theta - \frac{1}{2}\right) r'^2\, d^3r' \quad .$$

For an ellipsoid of uniform density the integral has a value

$$\frac{1}{5} M (b^2 - a^2)$$

and if $a = b + \Delta$ (for an oblate ellipsoid), $Q = \frac{2}{5}\frac{\Delta}{r_0}$.

The Sun of course is not of uniform density, and the relation between the quadrupole moment and the oblateness parameter Δ is more complicated. The effective potential near the surface of the sun is the gravitational potential (7.7.3) plus a term giving the centrifugal force

$$\varphi_{\text{effective}} = -\frac{GM}{r}\left\{1 - P_2(\cos\theta)\, Q\left(\frac{r_o}{r}\right)^2\right\} - \frac{1}{2}\omega^2 R^2 \qquad (7.7.4)$$

where R is a coordinate measured normal to the axis of rotation and ω is the angular velocity of the surface. To the extent to which the surface is maintained in equilibrium by pressure, the surface will follow the equipotential surfaces of Eq. (7.7.4).

Equating the values of φ_{eff} at the poles and on the equator gives

$$\frac{\Delta}{r_o} = \frac{3}{2} Q + \frac{1}{2}\frac{\omega^2 r_o^3}{GM} . \qquad (7.7.5)$$

In the case of a negligible quadrupole moment, the surface oblateness is determined by the $1/r$ field of the dense core and the centrifugal force and is

$$\frac{\Delta(Q=0)}{r_o} = \frac{1}{2}\frac{\omega^2 r_o^3}{GM_\odot} . \qquad (7.7.6)$$

The surface of the Sun rotates in approximately 26 days so

$$\frac{\Delta(Q=0)}{r_o} \simeq 10^{-5}$$

and this corresponds to negligible quadrupole moment. The opposite extreme is represented by taking a uniformly rotating Sun of uniform density and setting

$$Q = \frac{2}{5}\frac{\Delta}{r_o}$$

in Eq. (7.7.5). This yields an oblateness

$$\frac{\Delta(\text{uniform})}{r_o} = \frac{5}{4}\frac{\omega^2 r_o^3}{GM_\odot} = 2.5 \times 10^{-5}$$

and a quadrupole moment $Q = 10^{-5}$. Any uniformly rotating model of the Sun will be much closer to the former case than the latter. In the equatorial plane, $P_2(\cos\theta) = -\frac{1}{2}$ so setting $n = 2$ and $a = \frac{1}{2} Q r_o^2$ in Eq. (7.2.12) we expect a centennial precession for the perihelion of Mercury $\ll 1''$ from the solar quadrupole moment.

Dicke and Goldenberg measured an oblateness of 4.5×10^{-5}, corresponding to $Q = 2.5 \times 10^{-5}$ and a centennial perihelion advance of $3''$ for Mercury. Such an unexpectedly large quadrupole moment was attributed by Dicke to a rapidly rotating interior of the Sun, the outer layers being braked by interaction with the solar wind [5].

However, a more recent measurement yielded an oblateness parameter in accord with that expected for negligible quadrupole moment [6]. The same work found evidence for periods of excess equatorial brightness to which the results of Dicke and Goldenberg may be attributed. The present position is that there is no evidence that any significant part of the advance of the perihelion of Mercury is due to a solar quadrupole moment.

References

[1] S. Weinberg, 'Gravitation and Cosmology', (Wiley 1972), Section 9.5.

[2] Further details may be found in
S. Weinberg, 'Gravitation and Cosmology', (Wiley 1972), Section 8.6;
C.W. Misner, K.S. Thorne, J.A. Wheeler, 'Gravitation', (Freeman 1973) Section 40.5.

[3] R.A. Hulse and J.H. Taylor, Ap. J. Lett., 195, L51 (1975).
For a discussion of the relativistic effects in this system, see:
L.W. Esposito and E.R. Harrison, Ap. J. Lett., 196, L1 (1975);
C.M. Will, Ap. J. Lett., 196, L3 (1975).

[4] R.H. Dicke and H.M. Goldenberg, Ap. J. Supp., 27, 131 (1974).

[5] R.H. Dicke, Annual Reviews of Astronomy and Astrophysics, 8, 297 (1970).

[6] H.A. Hill et al, Phys. Rev. Lett., 33, 1497 (1974).
H.A. Hill and R.T. Stebbins, Ap. J., 200, 471 (1975).

CHAPTER 8

GRAVITATIONAL WAVES

8.1 Introduction

The first step in the generalisation of the gravitational Poisson equation

$$\nabla^2 h = 4\pi G \rho \qquad (4.1.8)$$

yielded

$$\left(\nabla^2 - \frac{1}{c^2}\frac{\partial^2}{\partial t^2}\right) h = 4\pi G \rho$$

which at once led us to expect gravitational waves propagating with velocity c, solutions of the free field equation

$$\left(\nabla^2 - \frac{1}{c^2}\frac{\partial^2}{\partial t^2}\right) h = 0$$

or

$$\Box\, h_{\mu\nu} = 0$$

for a tensor field. This may be compared with the electromagnetic equation

$$\Box\, A_\mu = 0 \,.$$

In the electromagnetic case the source of A_μ is the current $\mathcal{J}_\mu$ and we have (with $c = 1$)

$$\Box\, A_\mu = -4\pi \mathcal{J}_\mu$$

with the particular integral solution

$$A_\mu(\underline{r}, t) = \int \frac{d^3r' \, \mathcal{J}_\mu(\underline{r}', t - |\underline{r} - \underline{r}'|)}{|\underline{r} - \underline{r}'|} ,$$

the familiar retarded potential.

Our spin 2 tensor theory has

$$\Box\, h_{\mu\nu} = -8\pi G \left\{ \mathfrak{J}_{\mu\nu} - \tfrac{1}{2}\delta_{\mu\nu}\mathfrak{J}_{\sigma\sigma} \right\} = -8\pi G\, \bar{\mathfrak{J}}_{\mu\nu}$$

with the retarded potential

$$h_{\mu\nu}(\underline{r}, t) = 2G \int \frac{\bar{\mathfrak{J}}_{\mu\nu}(\underline{r}', t - |\underline{r} - \underline{r}'|)\, d^3r'}{|\underline{r} - \underline{r}'|} . \qquad (8.1.1)$$

which of course contains a radiative component with derivatives of the potential falling off as $1/r$.

8.2 Transverse nature of the waves

Because of the conservation law

$$\frac{\partial \mathcal{J}_\mu}{\partial x_\mu} = 0$$

we inferred for electromagnetism the Lorentz condition

$$\frac{\partial A_\mu}{\partial x_\mu} = 0 .$$

For gravitation we have

$$\frac{\partial \mathfrak{T}_{\mu\nu}}{\partial x_\mu} = 0$$

and so expect an analogous condition. But

$$h_{\mu\nu} \sim \bar{\mathfrak{T}}_{\mu\nu} = \left\{\mathfrak{T}_{\mu\nu} - \tfrac{1}{2}\delta_{\mu\nu}\mathfrak{T}_{\sigma\sigma}\right\} \qquad (8.2.1)$$

and so the analogous condition will NOT be

$$\frac{\partial h_{\mu\nu}}{\partial x_\mu} = 0$$

but rather

$$\frac{\partial \bar{h}_{\mu\nu}}{\partial x_\mu} = 0 \qquad (8.2.2)$$

where

$$\bar{h}_{\mu\nu} = h_{\mu\nu} - \tfrac{1}{2}\delta_{\mu\nu}h_{\sigma\sigma} .$$

Since

$$h_{\mu\nu} \sim \mathfrak{T}_{\mu\nu} - \tfrac{1}{2}\delta_{\mu\nu}\mathfrak{T}_{\sigma\sigma}$$

$$h_{\alpha\alpha} \sim \mathfrak{T}_{\alpha\alpha} - \tfrac{1}{2}\delta_{\alpha\alpha}\mathfrak{T}_{\beta\beta} ,$$

$$\bar{h}_{\mu\nu} = h_{\mu\nu} - \tfrac{1}{2}\delta_{\mu\nu}h_{\alpha\alpha} \sim \mathfrak{T}_{\mu\nu} - \tfrac{1}{2}\delta_{\mu\nu}\mathfrak{T}_{\sigma\sigma} - \tfrac{1}{2}\delta_{\mu\nu}\mathfrak{T}_{\alpha\alpha} + \tfrac{1}{4}\delta_{\mu\nu}\delta_{\alpha\alpha}\mathfrak{T}_{\beta\beta}$$

$$= \mathfrak{T}_{\mu\nu} - \tfrac{1}{2}\delta_{\mu\nu}\left\{\mathfrak{T}_{\sigma\sigma} + \mathfrak{T}_{\alpha\alpha} - \tfrac{1}{2}\delta_{\alpha\alpha}\mathfrak{T}_{\beta\beta}\right\}.$$

In these equations σ, α and β are dummy variables (summed over) and $\delta_{\alpha\alpha} = 4$ (that is, $\delta_{11} + \delta_{22} + \delta_{33} + \delta_{44}$) so the curly bracket sums to zero and we have

$$\bar{h}_{\mu\nu} = h_{\mu\nu} - \tfrac{1}{2}\delta_{\mu\nu}h_{\alpha\alpha} \sim \mathfrak{T}_{\mu\nu} \qquad (8.2.3)$$

and hence infer

$$\frac{\partial \bar{h}_{\mu\nu}}{\partial x_\mu} = 0 . \qquad (8.2.3a)$$

Now we know that the Lorentz condition $\partial A_\mu/\partial x_\mu = 0$ is related to the transverse nature of electromagnetic waves, so we might expect that Eq.(8.2.3a) will tell us something about the transversality of gravitational waves. In the electromagnetic case we have a four-vector potential A_μ and the Lorentz condition $\partial A_\mu/\partial x_\mu = 0$. At first sight this suggests three independent polarisations for an electromagnetic wave — and we know there are only two (our martians of Chapter 3 probably found this out too). But consider a plane wave :

$$A_\mu = a_\mu\, e^{i(\underline{k}.\underline{x} - \omega t)} = a_\mu\, e^{ik_\nu x_\nu} \qquad (8.2.4)$$

where a_μ is a constant 4-vector.

$$\Box A_\mu = 0 \qquad \text{yields} \quad k_\nu k_\nu = 0$$

and

$$\frac{\partial A_\mu}{\partial x_\mu} = 0 \qquad \text{yields} \quad k_\mu a_\mu = 0 \, .$$

But if $a_\mu \rightarrow a_\mu + bk_\mu$ these conditions are still satisfied and so without loss of generality we can replace a_μ by a'_μ where $a'_\mu = a_\mu + bk_\mu$ with b an arbitrary scalar. For a wave travelling in the z direction, the condition $k_\mu a_\mu = 0$ yields $a_4 = -ia_3$. Then choosing $a_3 + bk_3 = 0$ we have

$$a'_1 = a_1 \qquad a'_3 = 0$$

$$a'_2 = a_2 \qquad a'_4 = 0$$

and are left with only two polarisations. Since $a_\mu \rightarrow a_\mu + bk_\mu$ corresponds to setting

$$A_\mu \rightarrow A_\mu + \frac{\partial B}{\partial x_\mu} \, , \text{ with } B = -ib\, e^{ik_\nu x_\nu}$$

the equations of motion, determined by the fields

$$\underline{E} = -\left\{\underline{\nabla}\varphi + \frac{\partial \underline{A}}{\partial t}\right\}$$

$$\underline{B} = \underline{\nabla} \times \underline{A}$$

do not notice the transformation $a_\mu \rightarrow a_\mu + bk_\mu$ and indeed only two polarisations are physically significant for a plane wave, and hence any wave.

The potential transformation $a_\mu \rightarrow a_\mu + bk_\mu$ is called a gauge transformation, [1], and is quite evidently intimately connected with the conservation of the source of the electromagnetic field.

With

$$a_3 = a_4 = 0 \; , \quad \underline{E} = -\frac{\partial \underline{A}}{\partial t}$$

and electrons exposed to the wave vibrate in the x direction if driven by A_1 only, in the y direction if driven by A_2 only. In general of course the two components of the polarisation vector (a_1 , a_2) may have arbitrary magnitudes and relative phase and the familiar range of polarisation effects obtains. The vibration pattern essentially repeats under a rotation through 180° about the beam (with a change of phase of π) and we talk about spin 1 photons with two polarisation states $\hat{a}_1 , \hat{a}_2$ or two helicity states $\frac{1}{\sqrt{2}}(\hat{a}_1 \pm i\,\hat{a}_2)$.

The condition $\partial \bar{h}_{\mu\nu}/\partial x_\mu = 0$ results in a similar behaviour in the case of gravity, and in fact this behaviour motivates the choice of the tensor theory rather than the Newtonian theory.

Consider a plane wave

$$h_{\mu\nu} = \epsilon_{\mu\nu}\, e^{ik_\lambda x_\lambda} \; . \qquad (8.2.5)$$

In free space $\Box\, h_{\mu\nu} = 0$ so that

$$k_\lambda k_\lambda = 0 \; . \qquad (8.2.6)$$

The condition

$$\frac{\partial}{\partial x_\mu}\left\{h_{\mu\nu} - \tfrac{1}{2}\delta_{\mu\nu} h_{\sigma\sigma}\right\} = 0$$

yields

$$k_\mu \epsilon_{\mu\nu} - \tfrac{1}{2}\epsilon_{\sigma\sigma} k_\nu = 0 \; . \qquad (8.2.7)$$

If we now replace $\epsilon_{\mu\nu}$ by $\epsilon'_{\mu\nu} = \epsilon_{\mu\nu} + e_\mu k_\nu + e_\nu k_\mu$ (analogous to replacing a_μ by $a'_\mu = a_\mu + bk_\mu$ except that $\epsilon_{\mu\nu}$ has two indices and a_μ has one) we find

$$k_\mu \epsilon'_{\mu\nu} - \tfrac{1}{2}\epsilon'_{\sigma\sigma} k_\nu = \underline{k_\mu \epsilon_{\mu\nu} - \tfrac{1}{2}\epsilon_{\sigma\sigma} k_\nu} + k_\mu e_\mu k_\nu$$

$$+ \underline{e_\nu k_\mu k_\mu} - \tfrac{1}{2} k_\nu e_\sigma k_\sigma - \tfrac{1}{2} k_\nu e_\sigma k_\sigma \equiv 0$$

where the underlined pieces are zero by Eqs. (8.2.6) and (8.2.7).

We can choose four quantities e_μ arbitrarily and so go from ten independent components of $h_{\mu\nu}$ through Eq. (8.2.3a) to six independent components, and through the gauge invariance to two independent polarisations in a plane wave, and hence any wave. These are conventionally chosen to be h_{11} and h_{12} for a wave travelling in the $x_3(z)$ direction. This can be worked out quite easily.

The condition (8.2.7) $k_\mu \epsilon_{\mu\nu} - \frac{1}{2}\epsilon_{\sigma\sigma}k_\nu = 0$, gives

$$k_3\,\epsilon_{3\nu} + i\omega\,\epsilon_{4\nu} - \tfrac{1}{2}\left(\epsilon_{11} + \epsilon_{22} + \epsilon_{33} + \epsilon_{44}\right)k_\nu = 0$$

(since $k_1 = k_2 = 0$) .

$$\left.\begin{aligned}
\nu = 1 :&\quad k_3\,\epsilon_{31} = -\,i\omega\,\epsilon_{41} \quad \text{or} \quad \epsilon_{31} = -\,i\,\epsilon_{41} \\
\nu = 2 :&\quad k_3\,\epsilon_{32} = -\,i\omega\,\epsilon_{42} \qquad\quad \epsilon_{32} = -\,i\,\epsilon_{42} \\
\nu = 3 :&\quad \tfrac{1}{2}k_3(\epsilon_{33} - \epsilon_{11} - \epsilon_{22} - \epsilon_{44}) = -\,i\omega\,\epsilon_{43} \\
&\quad \text{or} \quad \epsilon_{33} - \epsilon_{11} - \epsilon_{22} - \epsilon_{44} = -\,2i\,\epsilon_{43} \\
\nu = 4 :&\quad k_3\,\epsilon_{34} + i\omega\,\epsilon_{44} - \tfrac{1}{2}i\omega\,(\epsilon_{11} + \epsilon_{22} + \epsilon_{33} + \epsilon_{44}) = 0 \\
&\quad \text{or} \quad (\epsilon_{11} + \epsilon_{22} + \epsilon_{33} - \epsilon_{44}) = -\,2i\,\epsilon_{34}
\end{aligned}\right\} \qquad (8.2.7a)$$

We may rewrite the last pair of equations in the form

$$\epsilon_{11} + \epsilon_{22} = 0 \quad ; \quad \epsilon_{33} - \epsilon_{44} = -\,2i\,\epsilon_{43} \qquad (8.2.7b)$$

These equations take ten components down to six independent components, and still hold when we replace $\epsilon_{\mu\nu}$ with $\epsilon'_{\mu\nu}$

$$\epsilon'_{\mu\nu} = \epsilon_{\mu\nu} + e_\mu k_\nu + e_\nu k_\mu \,.$$

Remembering that only k_3 and k_4 are non-zero, we find

$$\epsilon'_{11} = \epsilon_{11} \qquad \epsilon'_{22} = \epsilon_{22} \qquad \epsilon'_{12} = \epsilon_{12} \,.$$

Then

$$\left.\begin{aligned}
\epsilon'_{31} &= \epsilon_{31} + e_1k_3 & \epsilon'_{41} &= \epsilon_{41} + i\,\omega\, e_1 \\
\epsilon'_{32} &= \epsilon_{32} + e_2k_3 & \epsilon'_{42} &= \epsilon_{42} + i\,\omega\, e_2 \\
\epsilon'_{33} &= \epsilon_{33} + 2e_3k_3 & \epsilon'_{43} &= \epsilon_{43} + i\,\omega\, e_3 + k_3e_4 \\
\epsilon'_{34} &= \epsilon_{34} + e_4k_3 + i\,\omega e_3 & \epsilon'_{44} &= \epsilon_{44} + 2i\omega\, e_4
\end{aligned}\right\} \qquad (8.2.8)$$

Choose e_1 , e_2 , e_3 and e_4 to make the first column of (8.2.8) vanish. These are our extra four constraints. With e_1 and e_2 thus determined, the first two rows of Eq. (8.2.7a) yield

$$\epsilon'_{41} = \epsilon'_{42} = 0 \,.$$

With ϵ'_{34} set equal to zero by appropriate choice of e_3 and e_4 , ϵ'_{43} is automatically equal to zero. The second of Eq. (8.2.7b) then yields $\epsilon'_{44} = 0$.

The only surviving terms are :

$$h_{11} = - h_{22} \quad , \quad h_{12} = h_{21} \; .$$

We thus have once more two independent polarisations which correspond to spin 2, for the basic vibration pattern repeats after a rotation of 90° rather than 180°. This is enormously important, because we know from special relativity that a massless particle has only two possible spin states, along and against the direction of motion, [2]. The photon is a massless spin 1 particle and now we infer the existence of the graviton, a massless spin 2 particle.

In the half-tensor (Newtonian) theory, we have from source conservation

$$\frac{\partial h_{\mu\nu}}{\partial x_{\mu}} = 0 \quad \text{or} \quad k_{\mu}\epsilon_{\mu\nu} = 0 \; .$$

If we set

$$\epsilon'_{\mu\nu} = \epsilon_{\mu\nu} + k_{\mu}e_{\nu} + k_{\nu}e_{\mu}$$

we find

$$k_{\mu}\epsilon'_{\mu\nu} = \underline{k_{\mu}\epsilon_{\mu\nu}} + \underline{k_{\mu}k_{\mu}e_{\nu}} + k_{\mu}k_{\nu}e_{\mu} \; .$$

The condition $k_{\mu}e_{\mu} = 0$ is not automatically satisfied by any e_{μ}. Only three of the four quantities e_{μ} are independent and so instead of eight constraints on ten independent variables giving two polarisations we have seven constraints on ten independent variables giving three polarisations. The half-tensor (Newtonian) theory is thus a mixture of the tensor (spin 2) theory and the scalar (spin zero) theory, and in our apparently artificial construction of the spin 2 theory we were in fact subtracting off the scalar piece in the half-tensor theory. On the grounds of simplicity and elegance we would prefer the theory of gravitation to correspond to spin 2 or spin 0 rather than a mixture of the two, and in this framework the deflection of light by the Sun of twice the Newtonian value at last appears as an entirely natural result.

8.3 Physical effects of gravitational waves

In order to understand the content of the two physical polarisations we need to study the physical effects of the radiation. If a train of gravitational waves passes through a laboratory, what sort of physical effect does it produce ? The first thing that must be realised is that the local gravitational field is not directly detectable because we must look at the relative motion of at least two objects to detect an effect. Electrons in wires are sensitive to the electric field of electromagnetic radiation because in such a field the electrons go one way and the positive lattice goes the other. But everything accelerates the same way in the same gravitational field and so we need two separated objects, with their relative motion determined not by the first derivatives of the potential but by the second derivatives — the tidal forces.

We have therefore to take as our primordial gravitational antenna two (point-like) masses separated in space and measure their separation. Let us calculate the effect on this separation of a gravitational wave characterised by $(h_{11}, h_{22} = -h_{11})$ or $(h_{12}, h_{21} = h_{12})$.

We return to our equations of motion (5.4.4) developed originally for solar system problems :

$$\frac{dp_\rho}{dt} = m\tilde{v}_\mu \tilde{v}_\nu \frac{\partial h_{\mu\nu}}{\partial x_\rho} - m\tilde{v}_\mu \tilde{v}_\nu \frac{\partial h_{\rho\nu}}{\partial x_\mu} \quad . \quad (5.4.4) \qquad (8.3.1)$$

At this point we encounter what is at first sight a serious difficulty. A mass at rest has all elements of $T_{\mu\nu}$ equal to zero except for T_{44}. Our wave however is represented only by $h_{11}, h_{22}, h_{12}, h_{21}$ and therefore impinging on a mass at rest it gives

$$\frac{dp_\rho}{dt} = 0 \quad . \qquad (8.3.2)$$

This however is as seen from the ideal coordinate frame we have established free of the effect of all gravitational fields. In this frame two masses initially at rest suffer no acceleration. But it is not clear that this is the case from the point of view of the observer with a measuring stick stretched between the two masses. Remember that in time independent potentials, the tensor, half tensor and scalar fields gave the same h_{44} and slowed down clocks by the same amount. However, the h_{ii} components had opposite sign for the tensor and scalar theories, respectively shrinking and expanding measuring sticks, while the h_{ii} components were absent in the half tensor theory and measuring sticks were unaffected. We guess therefore that it is the h_{ii} components that alter the lengths of measuring rods and that in the ideal frame the two masses do not move but rather the length of a measuring rod next to them fluctuates. The local observer however has no way of telling his measuring rod is fluctuating in length and so interprets his observations as meaning that the separation of the two masses is fluctuating due to a periodic tidal force exerted in the passage of a gravitational wave. This interpretation is indeed implicit in the more formal developments of section 6.5.

Take equation (8.3.1) and set $h_{\mu\nu} = h_{11}(x_3, x_4)$. Then we have

$$\frac{dp_\rho}{dt} = m v_1^2 \frac{\partial h_{11}}{\partial x_\rho} - m \tilde{v}_\mu v_1 \frac{\partial h_{11}}{\partial x_\mu} \delta_{\rho 1} \tag{8.3.3}$$

or

$$\frac{d\underline{p}}{dt} = m v_1^2 \underline{\nabla} h_{11} - m \underline{v}_1 \left\{ \underline{v} \cdot \underline{\nabla} h_{11} + \frac{\partial h_{11}}{\partial t} \right\}$$

$$\frac{dE}{dt} = - E v_1^2 \frac{\partial h_{11}}{\partial t} \tag{8.3.4}$$

for the motion of a particle of mass m. In interpreting these equations we must first find expressions for the momentum and energy. Consider motion in the x_1 direction. The first thing we notice is that dE/dt is a function of v_1^2 : E is not changed in the slow motion limit. This suggests writing

$$E = \frac{E_0}{\sqrt{1 - v_1^2\, k}}$$

when

$$\frac{dE}{dt} \simeq \frac{E_0}{(1 - v_1^2 k)^{\frac{3}{2}}} \left\{ v_1 \frac{dv_1}{dt} + v_1^2 \frac{dk}{dt} \right\}$$

Then to first order in small quantities,

$$\frac{v_1 \dfrac{dv_1}{dt} + v_1^2 \dfrac{dk}{dt}}{1 - v_1^2} = - v_1^2 \frac{\partial h_{11}}{\partial t} . \tag{8.3.5}$$

We also have

$$\frac{dp_3}{dt} = m\, v_1^2 \frac{\partial h_{11}}{\partial x_3} . \tag{8.3.6}$$

A light ray along the x_1 direction is therefore deflected in the x_3 direction by an amount depending on the gradient of h_{11}. In a wave picture this requires an acceleration in the x_1 direction. If we set

$$k E v_1 = p_1$$

then

$$E \frac{dv_1}{dt} = \frac{dp_1}{dt} - v_1 \frac{dE}{dt} - E v_1 \frac{dk}{dt} = \left\{ - v_1 \frac{\partial h_{11}}{\partial t} + v_1^3 \frac{\partial h_{11}}{\partial t} - v_1 \frac{\partial k}{\partial t} \right\} E$$

which on multiplying by v_1 yields (8.3.5) again.

The refractive index necessary to produce the deflection in the x_3 direction implied by Eq. (8.3.6) requires

$$c_1 = (1 - h_{11})$$

and hence

$$k = 1 + h_{11} .$$

For slow motion

$$\frac{dp_1}{dt} = - m\, v_1 \frac{dh_{11}}{dt}$$

$$\frac{dE}{dt} = 0 \qquad (8.3.7)$$

and

$$E = \frac{E_0}{\sqrt{1 - k^2\, v_1^2}} \quad , \quad p_1 = \frac{k\, E_0\, v_1}{\sqrt{1 - k^2\, v_1^2}} \; .$$

In the presence of other forces

$$\frac{dp_1}{dt} = - p_1 \frac{\partial h_{11}}{\partial t} + F_1$$

$$\frac{dE}{dt} = v_1 F_1 \qquad (8.3.8)$$

and following the arguments of section 5.5 we make the new definition

$$p_1 \to (1 + h_{11})\, p_1 = \frac{(1 + 2h_{11})\, E_0\, v_1}{\sqrt{1 - k^2\, v_1^2}} \; .$$

Thus the rest energy is unchanged, the velocity of light in the x_1 direction is $c_1 = 1 - h_{11}$ and the rest mass $m_{0_1} \to m_0(1 + 2h_{11})$. The quantity $m_{0_1} c_1^2$ is the locally defined rest energy and is indeed unchanged. Since energy is unaffected, atomic frequencies are unchanged and the change in the velocity of light therefore implies that a length x_1 is changed by a factor $1 - h_{11}$. As a final check, for one dimensional motion in a Bohr atom we write

$$\tfrac{1}{2} m\, v_1^2 - \frac{e^2}{x_1} = \text{constant}$$

and this is indeed satisfied to first order in h_{11} if $x_1 \to x_1(1 - h_{11})$ and $e^2 \to e^2/(1 + h_{11})$.

We have then a consistent picture in the spirit of section 5.5. The effect of gravitational waves on the separation of two masses, initially at rest, is to provide a relative acceleration

$$\frac{d^2 x_i}{dt^2} = x_j^0 \frac{\partial^2 h_{ij}}{\partial t^2} \qquad (8.3.9)$$

where x_j^0 is the separation in the j direction in the absence of the wave and all measurements are made locally.

8.4 Polarisation properties of gravitational waves

The two independent polarisations are given by $(h_{11} = - h_{22})$ and $(h_{12} = h_{21})$.

We consider a set of four particles, arranged at the corners of a square normal to the direction of propagation of the gravitational wave. This is the minimum set of particles necessary to bring out the general features, since we need two particles to detect any effect and each polarisation affects length measurements in two orthogonal directions simultaneously. We obtain the sequences shown in Fig. 8.4.1.

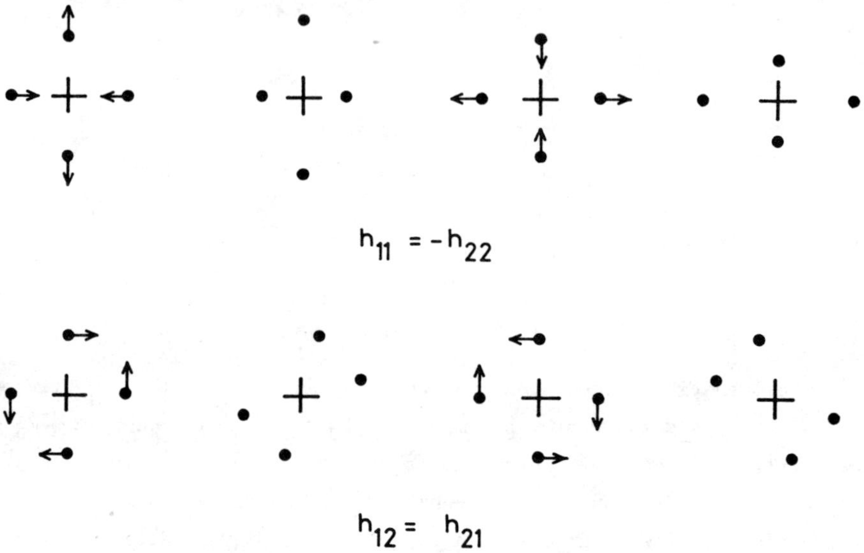

Fig.8.4.1 Positions and velocities of four test particles in a plane normal to the direction of a steady incident gravitational wave, as observed locally. The configurations are shown every quarter period

For a general wave we can combine these two polarisations with any magnitudes and relative phase. It is clear that these vibrations are quadrupole in character and that the elementary mode of vibration induced by a gravitational wave is quadrupole, just as the elementary mode of vibration induced by an electromagnetic wave is dipole (dipole oscillations cannot exist for gravity because there is no negative mass and hence no gravitational dipole).

8.5 Detection of gravitational waves

Our elementary gravitational antenna will therefore consist of a pair of masses M, connected by a spring if we want to make a resonant antenna, with some means of measuring the separation, (Fig. 8.5.1). The equation of such an antenna can then be written, for the case where the antenna is aligned with one of the principal stress axes ,

Fig. 8.5.1

$$\frac{d^2x}{dt^2} + \gamma\frac{dx}{dt} + \omega_o^2\, x \simeq \frac{\partial^2 h}{\partial t^2}\, x_o \qquad (8.5.1)$$

where x_o is the rest separation of the masses and x is the change in separation. At fixed frequency

$$x = -\frac{\omega^2 h\, x_o}{\omega_o^2 - \omega^2 - i\omega\gamma} \; . \qquad (8.5.2)$$

The spring constant controls the resonant frequency and the damping constant γ the Q value of the detector. A small value of γ gives a narrow band antenna, with a high Q, while large γ increases the bandwidth. If the frequency spread in a gravitational wave is $\ll \gamma$, the sensitivity of the detector to frequencies around ω_o depends inversely on γ. If the gravitational radiation is broad band, however, the energy absorbed is independent of γ.

We should also note that an antenna of this kind has directional properties. The quantity hx_o varies as $\cos 2\varphi$ or $\sin 2\varphi$ as we rotate the axis about the direction of propagation of the wave: measuring φ from the x-axis the former result is for (h_{11}, h_{22}) and the latter for (h_{12}, h_{21}). As we change the angle between the axis of the antenna and the direction of propagation, the tidal forces act on the projected separation L thus introducing a term in $\sin\theta$. The tidal force $hL\sin\theta$ acts normal to the direction of propagation, but longitudinal oscillations of the detector of amplitude ℓ are induced by the component of this force along the detector axis, introducing a second factor of $\sin\theta$. Therefore

$$\frac{d^2\ell}{dt^2} \sim hL\sin^2\theta\,\cos 2\varphi$$

for a polarisation (h_{11}, h_{22}). This is illustrated in Fig.8.5.2.

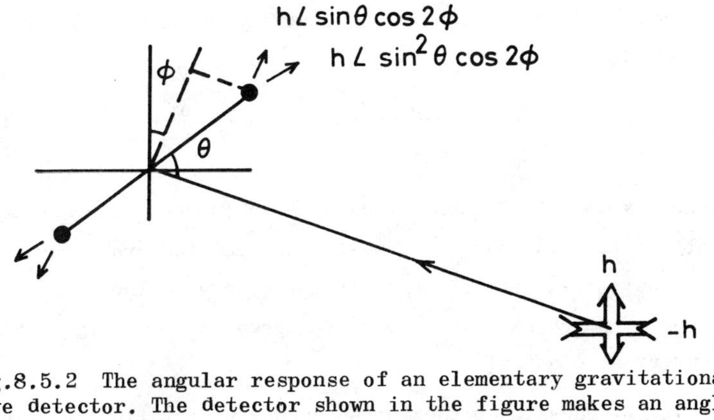

Fig.8.5.2 The angular response of an elementary gravitational wave detector. The detector shown in the figure makes an angle θ with the direction of propagation of the wave and has an azimuth measured from the h_{11} direction. The wave shown is characterised by $(h_{11}, h_{22} = -h_{11})$

The detection of gravitational waves is discussed in detail in the recent books by Misner, Thorne and Wheeler, and by Weinberg, [3]. Here we will only work out a few useful results. We take the case of a detector aligned with one of the stress axes of the gravitational wave, so that

$$x = -\frac{\omega^2 h x_o}{\omega_o^2 - \omega^2 - i\gamma\omega} \simeq \frac{\frac{1}{2}\omega_o h x_o}{\omega - \omega_o + i\frac{\gamma}{2}} . \tag{8.5.3}$$

Then

$$\frac{dx}{dt} \simeq \frac{\frac{1}{2} i \omega_o^2 h x_o}{\omega - \omega_o + i\frac{\gamma}{2}} . \tag{8.5.4}$$

(It should be remembered that in these equations h has the dimensions $\frac{GM}{Lc^2}$, that is, it is dimensionless.)

Locking ourselves firmly into the local system of coordinates defined by real measuring rods, the steady state vibrational energy of the antenna, in a monochromatic wave, is

$$E_v = 2\tfrac{1}{2}M\left(\frac{v_{max}}{2}\right)^2 = \frac{1}{16}\frac{M\omega_o^4 h_o^2 x_o^2}{(\omega-\omega_o)^2 + \frac{\gamma^2}{4}} . \tag{8.5.5}$$

The rate of dissipation of energy is E_v/τ, where $\tau = 1/\gamma$, so the rate of absorption **of** energy from the wave must be

$$\frac{1}{16}\frac{M\gamma\omega_o^4 h_o^2 x_o^2}{(\omega-\omega_o)^2 + \frac{\gamma^2}{4}} . \tag{8.5.6}$$

Averaging Eq. (8.5.6) over a broad frequency band gets rid of γ. Now the quantity

$$\frac{\omega^2 h^2 c^3}{G}$$

has the dimensions of energy flux. We rewrite the energy absorbed each second as

$$\frac{1}{16}\left[\frac{M\gamma\omega_o^2 x_o^2}{(\omega-\omega_o)^2 + \frac{\gamma^2}{4}}\frac{G}{c^3}\right]\left[\frac{\omega^2 h_o^2 c^3}{G}\right] \tag{8.5.7}$$

The left-hand box has the dimensions of an area and contains quantities characteristic only of the detector. It must represent the cross-section of the detector for the absorbtion of energy from the wave. The right-hand box has the dimensions of an energy flux and must represent the energy flux in the wave, down to a numerical constant.

We may tentatively guess that the instantaneous energy flux associated with (h_{11}, h_{22}) is :

$$\frac{c^3}{8\pi G}\left\{\left(\frac{\partial h_{11}}{\partial t}\right)^2 + \left(\frac{\partial h_{22}}{\partial t}\right)^2\right\} .$$

[This makes some sort of sense in the light of our knowledge of Newtonian gravity. The gravitational self-energy can be written as

$$-\frac{1}{8\pi G}\int\left(\underline{\nabla} h_{44}\right)^2 dV$$

with h having dimensions GM/R, and with h dimensionless as

$$-\frac{c^4}{8\pi G}\int\left(\underline{\nabla} h_{44}\right)^2 dV .]$$

With $-h_{22} = h_{11} = h_o\, e^{i(\underline{k}.\underline{r} - \omega t)}$, this gives us on time averaging an energy flux

$$S_G = \frac{c^3}{8\pi G} h_o^2\, \omega^2 \qquad (8.5.8)$$

with h dimensionless, or if h has dimensions GM/R,

$$S_G = \frac{c}{8\pi G} h_o^2\, \omega^2 . \qquad (8.5.8a)$$

With this tentative identification (note that we have attempted no proper derivation of the energy flux in gravitational radiation) we find the cross-section of the simple antenna we have been considering to be

$$\sigma = \frac{\pi}{2}\frac{G}{c^3}\,\frac{M\gamma\omega_o^2 x_o^2}{(\omega - \omega_o)^2 + \frac{\gamma^2}{4}} \qquad (8.5.9)$$

for a single frequency near resonance. In Eq. (8.5.9) we have suppressed the angular dependence of the cross-section, which may be inserted by multiplying by a factor $\sin^4\theta\cos^2 2\varphi$ if φ is measured from one of the stress axes of the radiation.

For a broad frequency band the power absorbed each second is thus

$$\sim \frac{\pi^2}{c^3} GM\omega_o^2\, x_o^2 F(\omega_o) \qquad (8.5.10)$$

where $F(\omega)\,d\omega$ is the energy flux in the incident radiation between ω and $\omega + d\omega$.

We are now in a position to consider briefly such experimental results as exist at present (1975) but to set the problem in perspective we first discuss the generation of gravitational waves and possible astrophysical sources of such radiation.

8.6 Generation of gravitational waves

We will consider briefly gravitational quadrupole radiation: much more detailed discussions may be found in Misner, Thorne and Wheeler and in Weinberg [4].

The solution (8.1.1) of the inhomogeneous wave equation is

$$h_{\mu\nu}(\underline{r}, t) = 2G \int \bar{\mathfrak{T}}_{\mu\nu} \frac{(\underline{r}', t - |\underline{r}-\underline{r}'|)}{|\underline{r}-\underline{r}'|} d^3r' .$$

For $r \gg r'$ we can write this as

$$h_{\mu\nu}(\underline{r}, t) \simeq \frac{2G}{r} \int \bar{\mathfrak{T}}_{\mu\nu} (\underline{r}', t - |\underline{r}-\underline{r}'|) \, d^3r' . \tag{8.6.1}$$

For a fixed frequency

$$\bar{\mathfrak{T}}_{\mu\nu}(\underline{r}', t) = \bar{\mathfrak{T}}_{\mu\nu}(\underline{r}') \, e^{-i\omega t}$$

and if the wavelength is large in comparison with the dimensions of the source the corrections to the phase of $h_{\mu\nu}(\underline{r}, t)$ due to variation of $\underline{r}'$ may be ignored, giving

$$h_{\mu\nu}(\underline{r}, t) \simeq \frac{2G}{r} e^{i(kr-\omega t)} \int \bar{\mathfrak{T}}_{\mu\nu}(\underline{r}') \, d^3r' . \tag{8.6.2}$$

We expect this approximation to correspond to quadrupole radiation, just as the equivalent approximation in electromagnetic theory corresponds to dipole radiation.

We have

$$h_{\mu\nu}(\underline{r}, t) \simeq \frac{2G}{r} e^{i(kr-\omega t)} \bar{T}_{\mu\nu}$$

where

$$\bar{T}_{\mu\nu} = \int \bar{\mathfrak{T}}_{\mu\nu} \, d^3r'$$

and we are interested in h_{11}, h_{22} and h_{12}.

$$T_{ij} = \int \mathfrak{T}_{ij}(\underline{r}') \, d^3r' = v_i v_j M \tag{8.6.3}$$

for a particle of mass M. For a single fixed frequency

$$T_{ij} \sim - \omega^2 x_i^0 x_j^0 M$$

and so

$$h_{ij}(\underline{r}, t) \approx - \frac{G}{r} e^{i(kr-\omega t)} \omega^2 Q_{ij}^0 \tag{8.6.4}$$

where Q_{ij}^0 is the amplitude of the quadrupole fluctuation.

If we work in units with h dimensionless, and write in c explicitly

$$h_{ij}(\underline{r}, t) \approx -\frac{G}{r} e^{i(kr - \omega t)} \frac{\omega^2}{c^4} Q^0_{ij} . \qquad (8.6.5)$$

The power radiated at the frequency ω is thus

$$P_\omega \approx G \frac{\omega^6}{c^5} |Q|^2 \qquad (8.6.6)$$

where Q is the quadrupole moment.

We can make a few numerical estimates. Suppose we have a close binary system with a period $T \sim 10$ hours, the mass of each member being $\sim M_\odot$. Then $Q \sim M_\odot R^2$, where

$$R \approx \left(\frac{GM_\odot T^2}{2\pi^2}\right)^{\frac{1}{3}} \approx 2 \times 10^{11} \text{ cm}$$

$$Q \approx 10^{56} \text{ gm cm}^2 , \quad \omega \sim 3 \times 10^{-5} \text{ s}^{-1} ,$$

and

$$P \approx 10^{30} \text{ erg s}^{-1}$$

which is $\approx 10^{-4}$ of the luminosity of the Sun. The energy of orbital motion is $\approx 10^{46}$ ergs, and so the decay time due to radiation of gravitational waves is $\approx 10^{16}$ sec $\approx 10^9$ years. This is a perfectly realistic example (the system containing the pulsar PSR 1913 + 16 has very much these characteristics).

As a more extreme and less realistic system suppose there to be a pair of neutron stars, masses $\sim M_\odot$ but separated by only $\approx 10^7$ cm . The period would be $\approx 10^{-2}$ sec, $Q \approx 10^{47}$ gm cm^2 and the power radiated $\approx 10^{46}$ erg s^{-1}. The energy in orbital motion would be $\approx 10^{53}$ erg and the corresponding decay time ≈ 1 year. (This is only one of the reasons why this example is unrealistic.)

Finally, suppose we have a neutron star rotating at a frequency $\omega \approx 10^4$ s^{-1} . This kind of frequency results if you consider a star with the mass, radius and rotation period of the Sun collapsing to a radius $\sim 10^6$ cm with internal conservation of angular momentum. A similar result is obtained by extrapolating back the slowdown rate of the pulsar NP 0532, in the Crab nebula, to its birth, 1054 A.D. (The present pulse period is 33 ms.) Then

$$P \approx 10^{-36} |Q|^2 \text{ erg s}^{-1}$$

and $Q \sim 10^{45}$ e where e is an ellipticity factor measuring a departure from rotational symmetry.

$$P \approx 10^{54} e^2 \text{ erg s}^{-1}$$

and the rotational energy is $\approx 10^{53}$ ergs. The conditions described are close to breakup and so in the early stages of neutron star formation one might get

$\sim 10^{53}$ ergs s^{-1} for around 1 sec. If the neutron star avoids breakup and settles down with $e \approx 10^{-4}$, then

$$P \approx 10^{46} \text{ erg s}^{-1}$$

and the slowdown period is ≈ 1 year.

We may compare these figures with the total energy released in gravitational collapse of a star like the Sun to nuclear densities where $R \approx 10^6$ cm. This is $\approx 10^{53}$ ergs. The conversion of $1\,M_\odot$ into gravitational energy would yield $\approx 10^{54}$ ergs.

It would seem that we have no right to expect sources of gravitational radiation producing more than $\approx 10^{54}$ erg over ≈ 1 sec or more than $\approx 10^{46}$ erg s^{-1} over ≈ 1 year, at least if we restrict ourselves to simple stellar systems.

8.7 Attempts to detect gravitational waves

Suppose we have a burst of gravitational radiation lasting ~ 1 sec. The energy absorbed in a simple resonant antenna is

$$P_I = \int P\,dt \approx 10^{-30}\,G M \omega_o^2 x_o^2 F(\omega_o) \qquad (8.7.1)$$

where M, ω_o and x_o are characteristics of the antenna.

The first attempts to detect gravitational radiation were made by Weber [5] who employed as detectors several aluminium cylinders in Maryland and a second array at Argonne, some 1000 km away.

These cylinders have a mass of 1.4×10^6 gm, a length of 153 cm and the natural frequency of the lowest longitudinal vibration mode is 1660 c.p.s. For such a cylinder

$$P_I \approx 10^{-19}\,F(\omega_o) \text{ erg}\,.$$

If we take as a rough guide to the signal discernable above noise

$$P_I \gtrsim kT \approx 5 \times 10^{-14} \text{ erg}$$

for a noise temperature of 300° K, then for a pulse to be detectable

$$F(\omega_o) \gtrsim 10^4 \text{ erg cm}^{-2}\text{ s}^{-1} \text{ per cycle.}$$

The total energy flux would have to be $\gtrsim 10^7$ erg cm^{-2} s^{-1} over the period of the pulse. If the source were located at the galactic centre and radiating isotropically, this means an energy release $\gtrsim 10^{53}$ erg to make up the pulse; not impossible, but nudging the upper limits we worked out in the previous section.

Suppose on the other hand 1660 c.p.s. is essentially on frequency for the rotation of a collapsed deformed star. The power picked up each second then depends on the Q value of a resonant antenna: for Weber's cylinders the constant $\gamma \approx 0.1\ s^{-1}$.

The energy absorbed each second is $\approx (10^{-19}/\gamma)$ F erg, where F is the incident flux. If the steady state vibrational energy $\approx kT$ for detection of this signal, and the natural decay time of vibrations in the cylinder is ~ 10 sec, then $F \approx 10^4$ erg cm^{-2} s^{-1}. This corresponds to an isotropic source at the centre of the galaxy radiating $\approx 10^{48}$ erg s^{-1}.

We may conclude that if we can pick up induced vibrations at this noise level, catastrophic events near the centre of the galaxy are detectable, but steady single frequency signals of the kind that might be produced over the first few years after neutron star formation would have to have a fairly local source, say within 5,000 light years.

It is salutory to calculate the amplitude of vibration corresponding to an energy $10^{-14} - 10^{15}$ ergs.

$$10^{-14} - 10^{-15} \approx M\omega_0^2 (\Delta x)^2 \approx 10^{12} (\Delta x)^2$$

so $\Delta x \approx 10^{-13} - 10^{-14}$ cm. This is less than the diameter of an atomic nucleus. The corresponding strain is $\sim 10^{-15} - 10^{-16}$. There are in fact several ways of measuring displacements and strains of these sizes. Weber's detectors utilize the piezo-electric effect, with piezo-electric transducers glued to the sides of the aluminium cylinder. In other similar antennae, two half cylinders are bonded together via a slab of piezo-electric material. The piezo-electric first used by Weber was quartz, which develops $0.16\,x$ Coulomb surface charge per square metre for a strain x. This corresponds to an electric dipole moment per cubic centimetre of 10^{-2} e-cm for a strain of 10^{-16}: a dipole moment per molecule $\sim 10^{-25}$ e-cm. Such a dipole moment means a separation of unit electron charges of $\approx 10^{-17}$ atomic diameters, a fractional displacement of charge of the same order as the strain.

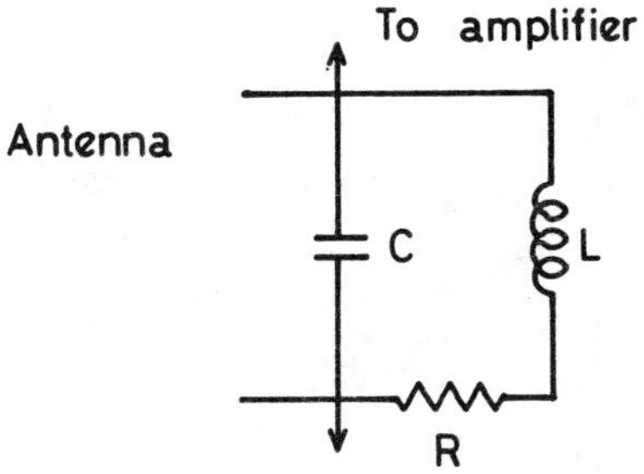

Fig.8.7.1

In Weber's later work he used ceramic barium titanate transducers: 44 of dimensions $5 \times 5 \times 1.2$ cm^3 bonded to the equator of the cylinder, where the strain is greatest [6]. With a surface area of ≈ 200 cm^2 the total charge produced for a strain of 10^{-16} is $\approx 10^{-19}$ Coulomb, one electron charge. This fluctuating charge was measured from the voltage developed across a high impedance: Weber [7] gives the equivalent circuit of Fig.8.7.1, with $C = 6\times10^{-9}$ f, $L = 1.6$ henry, $R = 0.1$ ohms. The impedance of this parallel tuned circuit is

$$Z = \frac{R + i\omega L}{Ri\omega C - \omega^2 LC + 1}$$

which is approximately L/RC on resonance, $\omega_R = \frac{1}{LC}$. With the above values, $\omega_R = 10^4$ c.p.s., $\nu_R = 1660$ c.p.s. and $Z_R = 3\times10^9$ ohms.

Since $V = IZ = i\omega q Z$ the amplitude of the voltage developed is $\approx 10^{-6}$ volts for a charge of 10^{-19} Coulomb. This fluctuating voltage is then fed through a low noise amplifier.

Another technique that has been used is to sense the relative displacement of the ends of a cylinder through changes in the capacitance of a parallel plate capacitor with one plate linked to each end. A relative displacement of the ends of amplitude 4×10^{-14} cm has been sensed as a radio frequency voltage of amplitude 4×10^{-7} volts, [8].

Displacements $\approx 10^{-13}$ cm have also been measured using lasers. Timing pulses is limited to ~ 10 cm accuracy, measurement of fringe shifts in optical interferometry to $\approx 10^{-6}$ cm, but by picking out the beat frequency between light reflected back over a fluctuating path length and light which has travelled a fixed path, very small amplitudes can be measured. A 5000 c.p.s. vibration of amplitude 3×10^{-13} cm has been measured in this way [9]. This was done with a laser beam, the two paths being arms of a Michelson interferometer. After recombining the split beam the amplitude is

$$A\left\{\sin\omega t + \sin(\omega t + \varphi)\right\} \tag{8.7.2}$$

where φ is the phase difference between the beams that traversed the active and passive arms of the interferometer.

$$\varphi = \frac{4\pi}{\lambda}x(t) + \varphi_0$$

where x is the instantaneous amplitude of the vibration. The recombined beam is detected photoelectrically and the output of the photodetector is proportional to the square of Eq. (8.7.2) which after averaging over the optical frequency vibrations is

$$2A^2\,\overline{\sin^2\omega t}\,[1 + \cos\varphi]\,.$$

For maximum sensitivity, choose $\varphi_0 = \pi/2$ when the beat signal is

$$\sim 2A^2\,\overline{\sin^2\omega t}\,\sin\left[\frac{4\pi}{\lambda}x(t)\right].$$

Let $x = x_0 \sin \omega' t$ and for small x_0/λ the beat signal is

$$\sim 2A^2 \overline{\sin^2 \omega t} \frac{4\pi}{\lambda} x_0 \sin \omega' t$$

and the amplitude of the beat signal yields x_0/λ. The limiting factor in the work of reference [9] was photon noise.

Weber [5,7] has reported coincidences between his Maryland and Argonne arrays at the rate of ~100 per year, apparently explicable only in terms of pulses of gravitational radiation. In 1973 he reported seven events/day above a level of kT/100, [10]. He has used the $\sin^4\theta$ dependence of the cross-section to get direction information, finding the rate to be correlated with sidereal time (but not with solar time). Weber claims a clear excess of pulses travelling along a line joining the solar system with the galactic centre [11] (with a simple quadrupole antenna it is not possible to tell which way such pulses are travelling).

These results are astonishing. Supernova explosions are observed from Earth in our galaxy every few hundred years. From observations of other galaxies a total rate of about 1 supernova in 30 years has been inferred. For isotropic radiation, 100 detectable pulses each year from the centre of the galaxy implies an energy output $> 10^{55}$ erg each year. This is greater than the energy radiated as starlight in the galaxy, and means the conversion into gravitational waves of $> 100\ M_\odot$ each year. The galaxy contains $\sim 10^{11}$ stars and has existed for $\sim 10^{10}$ years: this is a level at which galactic dynamics should be visibly affected, [12].

If Weber's interpretation of his data is correct, it would seem that either the radiation must be strongly beamed in the galactic plane or that activity in the core of the galaxy has only recently switched on. In either case the core of our galaxy must be a far more violent region than we had any prior reason to suspect.

A number of other detectors of gravitational radiation have been brought into operation over the last few years [8,13]. In no case has any independent experiment confirmed Weber's exciting observations, which accordingly must be regarded as far from proven.

At present gravitational wave detectors are probing into the region of sensitivity to strains $\sim 10^{-18}$ and the ultimate hope is to achieve a sensitivity sufficient to detect cataclysmic events occurring in the rich Virgo cluster of galaxies.

References

[1] See for example J.D. Jackson, 'Classical Electrodynamics', (Wiley 1962) p.181.

[2] See for example H. Muirhead, 'Special Relativity' (Macmillan 1974) Chapter 6.

[3] C.W. Misner, K.S. Thorne and J.A. Wheeler, 'Gravitation' (Freeman 1973) Ch. 37.
S. Weinberg, 'Gravitation and Cosmology', (Wiley 1972) Ch. 10.7.

[4] S. Weinberg, 'Gravitation and Cosmology', (Wiley 1972) Ch. 10.5.
C.W. Misner, K.S. Thorne amd J.A. Wheeler, 'Gravitation' (Freeman 1973) Ch. 36.

[5] J. Weber, Phys. Rev. Lett., 20, 1307 (1968).
Phys. Rev. Lett., 22, 1320 (1969).

[6] J. Weber, Lett. Nuovo Cimento, 4, 653 (1970).

[7] J. Weber, Physics Today, 21, (4), 34 (1968).

[8] V. Braginsky et al., JETP Lett., 16, 108 (1972).

[9] G.E. Moss et al., Appl. Optics, 10, 2495 (1971).

[10] J. Weber, Phys. Rev. Lett., 31, 779 (1973).

[11] J. Weber, Phys. Rev. Lett., 25, 180 (1970).

[12] G.B. Field et al., Comments on Astrophysics and Space Physics, 1, 187 (1969).

[13] J.L. Levine and R.L. Garwin, Phys. Rev. Lett., 31, 173 (1973).
Phys. Rev. Lett., 31, 176 (1973).
Phys. Rev. Lett., 33, 794 (1974).
J.A. Tyson, Phys. Rev. Lett., 31, 326 (1973).
R.W.P. Drever et al., Nature, 246, 340 (1973).
J. Hough et al., Nature, 254, 498 (1975).
D.H. Douglass et al., Phys. Rev. Lett., 35, 480 (1975),
H. Billing et al., Lett. Nuovo Cimento, 12, 111 (1975).
See also J. Logan, Physics Today, 26, (3), 266 (1973);
and Physics Today, 26, (10), 17 (1973).

CHAPTER 9

GRAVITATION AND THE GEOMETRY OF SPACETIME

9.1 Introduction

Throughout most of this book we have been working in the standard coordinate systems of special relativity. We set up the Lorentz covariant equations for the tensor gravitational field and calculated the deflection of light relative to such an Olympian reference frame, and the radar echo delay due to the interaction of electromagnetic waves with the gravitational field. The equations of motion that were forced on us required a redefinition of momentum in addition to the obvious redefinition of energy in the presence of gravitation, and we then discovered that a gravitational potential distorts physical measuring rods and slows down clocks, relative to our Olympian reference frame. In a freely falling laboratory of such a size that changes in the gravitational potential over the laboratory could be neglected, we found the velocity of light, as it would be measured, to be constant. We also found that the kinematic equations of special relativity recover their standard form when everything is measured in local terms. As a consequence the principle of equivalence holds (at least for light and relativistic kinematics) and the Olympian reference frame is locally unobservable.

In these circumstances we may either adopt the point of view that gravity bends measuring sticks and slows clocks in the spacetime of special relativity, or we may consider these instruments as embedded in a spacetime which is itself distorted by the presence of mass. From the former point of view it might prove advantageous to formulate the theory of gravitation in terms of the locally measured quantities rather than sitting in an unobservable reference frame: from the latter point of view it is the obvious way to proceed.

The principle of equivalence is put in as the starting point, in the following form. In a sufficiently small region of spacetime, the laws of physics in a freely falling laboratory are the laws of special relativity, including their numerical content. The laws of physics in any other reference frame — a reference frame locked to a gravitating body will usually prove most convenient — may be obtained by a coordinate transformation alone, and this transformation is linked to the gravitational fields.

The Lorentz transformations themselves link any two inertial frames and do not contain any gravitational attributes. In an accelerating frame inertial forces appear which are reflected in the transformations: the recipe is to transform from a given inertial frame to the instantaneously comoving frame. These inertial forces may be regarded as gravitational forces, but of a particular kind. There exists a transformation (back to the original inertial frame or any other) that cancels them completely, everywhere and everywhen. No such transformation exists for permanent gravitational fields. A spacecraft in free fall close to the earth has an acceleration $\sim g$, as measured in an earthbased system of coordinates. In a frame of reference falling with the spacecraft there are (locally) no gravitational effects. The two frames are linked by a coordinate transformation. But if one spacecraft is falling towards the Atlantic Ocean and another towards Siberia, the two associated freely falling frames are approaching each other at $\sim 20\ \mathrm{km\ s^{-1}}$. In one freely falling frame the other spacecraft appears to be accelerating at $\sim 2g$, due to the gravitational field of the earth. It is when permanent gravitational fields are present that we speak of a curved spacetime, for whatever global transformation of coordinates we make, we cannot globally get rid of gravitation.

9.2 The metric tensor and equations of motion in free fall

We shall be considering particle dynamics and the propagation of light in gravitational fields. Consider a freely falling particle as viewed from a local freely falling frame. The principle of unique acceleration tells us that any such particle is not accelerating in the local freely falling frame. Set up in such a frame the standard axes of special relativity $X_\mu = (x, y, z, ict)$ where μ runs from 1 to 4. Lack of acceleration is written as

$$\frac{d^2X_j}{dX_4^2} = 0 \tag{9.2.1}$$

where j runs from 1 to 3. The proper time elapsed in the rest frame of the particle is a Lorentz invariant given by

$$d\tau^2 = - dX_\mu dX_\mu \tag{9.2.2}$$

with $c = 1$. Because there is no acceleration an interval of proper time $\Delta\tau$ is strictly proportional to an interval of coordinate time Δt in a local freely falling frame, so Eq. (9.2.1) may be written in the form

$$\frac{d^2X_\mu}{d\tau^2} = 0 . \tag{9.2.3}$$

It is convenient to redefine the coordinates in a freely falling frame so that

$$\xi_\alpha = (ct , x , y , z)$$

when we have in place of Eq. (9.2.3)

$$\frac{d^2 \xi_\alpha}{d\tau^2} = 0 \tag{9.2.4}$$

with the invariant interval, which corresponds to proper time on the particle, given by

$$d\tau^2 = - \eta_{\alpha\beta} \, \xi_\alpha \, \xi_\beta \tag{9.2.5}$$

$$\begin{aligned} \eta_{\alpha\beta} &= 1 \qquad \alpha = \beta = 1,2,3 \\ &= -1 \qquad \alpha = \beta = 0 \\ &= 0 \qquad \alpha \neq \beta \end{aligned}$$

We now make a transformation to another reference frame in which the particle coordinates are x_μ, where μ now runs from 0 to 3. We do not at this stage attach any significance to this new frame of coordinates. The transformation might denote something as trivial as a redefinition of the coordinates in the original frame, for example $x_1 = \ln \xi_1$ if there was some point in doing this. (This sort of transformation is made whenever logarithmic graph paper is used.) The transformation might merely be to another frame related to the original by the Lorentz transformation, to an accelerating frame in the absence of permanent gravitational fields, or to a frame of reference locked to a gravitating object like the Sun. So far, it does not matter. The prescription that we follow is that the transformed physical laws are still valid and with the same numerical content. In particular, an invariant such as proper mass or an interval of proper time still has the same numerical value. Physical laws which maintain their form and numerical content under such general transformations are generally covariant (as opposed to merely Lorentz covariant) and this is the origin of the term general relativity.

Either the old set of coordinates ξ_α or the new set x_μ are sufficient to describe completely the motion of a test particle. We may therefore write

$$d\xi_\alpha = \frac{\partial \xi_\alpha}{\partial x_\mu} \, dx_\mu \, . \tag{9.2.6}$$

This is just an expansion in terms of partial differentials. Therefore

$$\begin{aligned} d\tau^2 &= - \eta_{\alpha\beta} \, d\xi_\alpha \, d\xi_\beta = - \eta_{\alpha\beta} \frac{\partial \xi_\alpha}{\partial x_\mu} \frac{\partial \xi_\beta}{\partial x_\nu} \, dx_\mu \, dx_\nu \\ &= - g_{\mu\nu} \, dx_\mu \, dx_\nu \end{aligned} \tag{9.2.7}$$

where

$$g_{\mu\nu} = \eta_{\alpha\beta} \frac{\partial \xi_\alpha}{\partial x_\mu} \frac{\partial \xi_\beta}{\partial x_\nu} \quad . \tag{9.2.8}$$

The quantity $g_{\mu\nu}$ is clearly symmetric. It is important to note that the quantity $d\tau^2$ is generally invariant, having the same value in either the ξ-frame or the x-frame. For example, in an inertial frame (a local freely falling frame) the invariant interval is zero for light. In an arbitrary coordinate system therefore the propagation of light must follow the equation

$$g_{\mu\nu}\, dx_\mu\, dx_\nu = 0$$

and this is a generally covariant equation. With $d\tau^2$ an invariant, we may transform Eq. (9.2.4) as follows

$$\begin{aligned}
\frac{d^2\xi_\alpha}{d\tau^2} &= \frac{d}{d\tau}\left\{\frac{d\xi_\alpha}{d\tau}\right\} = \frac{d}{d\tau}\left\{\frac{\partial \xi_\alpha}{\partial x_\mu} \frac{dx_\mu}{d\tau}\right\} \\
&= \frac{\partial \xi_\alpha}{\partial x_\mu} \frac{d^2 x_\mu}{d\tau^2} + \frac{dx_\mu}{d\tau} \frac{d}{d\tau}\left\{\frac{\partial \xi_\alpha}{\partial x_\mu}\right\} \\
&= \frac{\partial \xi_\alpha}{\partial x_\mu} \frac{d^2 x_\mu}{d\tau^2} + \frac{dx_\mu}{d\tau} \frac{dx_\nu}{d\tau} \frac{\partial^2 \xi_\alpha}{\partial x_\mu\, \partial x_\nu} = 0 \ .
\end{aligned}$$

This is again just an expansion in terms of partial differentials, remembering that neither ξ_α nor x_μ are explicit functions of τ : $\xi_\alpha = \xi_\alpha\,(x_\mu)$ or $x_\mu = x_\mu\,(\xi_\alpha)$.

We now have a set of four equations relating the four components of four-acceleration to the four-velocities and the partial differential coefficients. We may solve them simultaneously to obtain the four-acceleration. Multiply each of the four equations, labelled by the only index which is not dummy, α, by the quantity $\partial x_\lambda / \partial \xi_\alpha$ where λ has any value from 0 to 3 , and add all four equations. This operation is easily written using the convention of summation over repeated indices

$$\frac{\partial x_\lambda}{\partial \xi_\alpha} \frac{\partial \xi_\alpha}{\partial x_\mu} \frac{d^2 x_\mu}{d\tau^2} + \frac{dx_\mu}{d\tau} \frac{dx_\nu}{d\tau} \frac{\partial x_\lambda}{\partial \xi_\alpha} \frac{\partial^2 \xi_\alpha}{\partial x_\mu\, \partial x_\nu} = 0 \ . \tag{9.2.9}$$

Now note that

$$\frac{\partial x_\lambda}{\partial \xi_\alpha} \frac{\partial \xi_\alpha}{\partial x_\mu} = \frac{\partial x_\lambda}{\partial x_\mu} = \delta_{\lambda\mu}$$

because each coordinate in one system is a function of the four coordinates in the other system. Thus on summing over α we only retain the term for which $\mu = \lambda$ in the left-hand piece of Eq. (9.2.9). Then

$$\frac{\partial^2 x_\lambda}{d\tau^2} + \frac{\partial x_\lambda}{\partial \xi_\alpha}\frac{\partial^2 \xi_\alpha}{\partial x_\mu \partial x_\nu}\frac{dx_\mu}{d\tau}\frac{dx_\nu}{d\tau} = 0 \tag{9.2.10}$$

with

$$d\tau^2 = -g_{\mu\nu}\, dx_\mu\, dx_\nu .$$

The equation (9.2.10) contains a single coefficient composed of partial differentials and we write

$$\frac{d^2 x_\lambda}{d\tau^2} + \Gamma^\lambda_{\mu\nu}\frac{dx_\mu}{d\tau}\frac{dx_\nu}{d\tau} = 0 \tag{9.2.11}$$

where the coefficient $\Gamma^\lambda_{\mu\nu}$ is a function only of the coordinates (x_0, x_1, x_2, x_3).

Now we notice that

$$\frac{\partial g_{\mu\nu}}{\partial x_\lambda} = \frac{\partial}{\partial x_\lambda}\left\{\eta_{\alpha\beta}\frac{\partial \xi_\alpha}{\partial x_\mu}\frac{\partial \xi_\beta}{\partial x_\nu}\right\} = \eta_{\alpha\beta}\left\{\frac{\partial^2 \xi_\alpha}{\partial x_\lambda \partial x_\mu}\frac{\partial \xi_\beta}{\partial x_\nu} + \frac{\partial^2 \xi_\beta}{\partial x_\lambda \partial x_\nu}\frac{\partial \xi_\alpha}{\partial x_\mu}\right\} \tag{9.2.12}$$

and is clearly related to the Γ's. It looks as though the Γ's can be expressed in terms of the quantity $g_{\mu\nu}$ and its first derivatives and we can at once anticipate that the $g_{\mu\nu}$ can be interpreted as gravitational potentials and the Γ's as gravitational fields.

With

$$\frac{\partial x_\rho}{\partial \xi_\alpha}\frac{\partial^2 \xi_\alpha}{\partial x_\mu \partial x_\nu} = \Gamma^\rho_{\mu\nu}$$

we may multiply by $\partial \xi_\alpha / \partial x_\rho$ and summing over ρ on the left-hand side before attempting to sum over α we have

$$\frac{\partial \xi_\gamma}{\partial x_\rho}\frac{\partial x_\rho}{\partial \xi_\alpha} = \frac{\partial \xi_\gamma}{\partial \xi_\alpha} = \delta_{\alpha\gamma}$$

so that

$$\frac{\partial^2 \xi_\gamma}{\partial x_\mu \partial x_\nu} = \frac{\partial \xi_\gamma}{\partial x_\rho}\Gamma^\rho_{\mu\nu} .$$

Substituting in Eq. (9.2.12)

$$\frac{\partial g_{\mu\nu}}{\partial x_\lambda} = g_{\rho\nu}\Gamma^\rho_{\lambda\mu} + g_{\rho\mu}\Gamma^\rho_{\lambda\nu} . \tag{9.2.13}$$

The only dummy variable is ρ, so Eq. (9.2.13) is shorthand for a set of 64 equations relating the 64 first derivatives of $g_{\mu\nu}$ to the 64 quantities Γ. (Because $g_{\mu\nu}$ is symmetric with only ten independent components we actually have 40 equations relating 40 derivatives to 40 independent Γ's.) These equations may be shuffled around, [1], to yield

$$\Gamma^\sigma_{\lambda\mu} = \tfrac{1}{2} g^{\nu\sigma}\left\{\frac{\partial g_{\mu\nu}}{\partial x_\lambda} + \frac{\partial g_{\lambda\nu}}{\partial x_\mu} - \frac{\partial g_{\mu\lambda}}{\partial x_\nu}\right\} \tag{9.2.14}$$

where the quantity $g^{\nu\sigma}$ is the inverse of $g_{\sigma\nu}$, defined by $g^{\nu\sigma}g_{\sigma\rho} = \delta_{\nu\rho}$. We thus have

$$\frac{d^2x_\lambda}{d\tau^2} = -\tfrac{1}{2}g^{\rho\lambda}\left\{\frac{\partial g_{\mu\rho}}{\partial x_\nu} + \frac{\partial g_{\nu\rho}}{\partial x_\mu} - \frac{\partial g_{\mu\nu}}{\partial x_\rho}\right\}\frac{\partial x_\mu}{d\tau}\frac{\partial x_\nu}{d\tau} \tag{9.2.15}$$

or

$$g_{\rho\lambda}\frac{d^2x_\lambda}{d\tau^2} = -\tfrac{1}{2}\left\{\frac{\partial g_{\mu\rho}}{\partial x_\nu} + \frac{\partial g_{\nu\rho}}{\partial x_\mu} - \frac{\partial g_{\mu\nu}}{\partial x_\rho}\right\}\frac{dx_\mu}{d\tau}\frac{dx_\nu}{d\tau} . \tag{9.2.16}$$

At this stage we may throw away the partial differentials involving the ξ's, which are only locally defined, and the equations of free fall are expressed in terms of the purely geometric quantity $g_{\mu\nu}$ and its first derivatives. The quantity $g_{\mu\nu}$ is called the metric tensor and the quantities $\Gamma^\lambda_{\mu\nu}$ the affine connections.

Now Eq. (6.4.11) may be rewritten with the same indices as (9.2.16) for easier comparison. Eq. (6.4.11) becomes

$$(\delta_{\rho\lambda} + 2h_{\rho\lambda})\frac{d^2x_\lambda}{d\tau^2} = \left\{\frac{\partial h_{\mu\nu}}{\partial x_\rho} - \frac{\partial h_{\mu\rho}}{\partial x_\nu} - \frac{\partial h_{\nu\rho}}{\partial x_\mu}\right\}\frac{dx_\mu}{d\tau}\frac{dx_\nu}{d\tau} . \tag{9.2.17}$$

Equation (9.2.16) is written in terms of purely geometric quantities and bears a great resemblance to Eq. (9.2.17) written in terms of gravitational potentials. If we were to set

$$g_{\rho\lambda} = \delta_{\rho\lambda} + 2h_{\rho\lambda}$$

the equations would be identical. It must be remembered that Eq. (9.2.16) follows the convention

$$x_\mu = (ct, x, y, z)$$

while Eq. (9.2.17) follows the convention of special relativity

$$x_\mu = (X, Y, Z, icT) .$$

We could clearly have followed the latter convention in the preceding work, in which case in the absence of gravitational fields $g_{\mu\nu} \to \delta_{\mu\nu}$ rather than $g_{\mu\nu} \to \eta_{\mu\nu}$.

These two equations are therefore identical when written using the same conventions and in the ict convention identifying

$$g_{\mu\nu} = \delta_{\mu\nu} + 2h_{\mu\nu} .$$

(We could equally well rewrite Eq. (6.5.11) and related equations in the $x_0 = ct$ convention.) The terms of the metric tensor thus represent gravitational potentials and their derivatives gravitational force fields: the work of this section and that of section 6.5 tie nicely together. We have however no recipe so far for finding the $g_{\mu\nu}$ corresponding to a permanent gravitational field.

9.3 Concerning the field equations

In Chapter 4 we were faced with the problem of finding the gravitational potentials within the framework of special relativity. We solved this problem by searching for Lorentz covariant equations with the right Newtonian limit, and had then to determine the final form by appealing to the measurement of the deflection of light by the Sun. The tensor theory we constructed turned out in Chapter 8 to have a gauge invariance property and correspond to spin 2 gravitations. The Einstein field equations for the $g_{\mu\nu}$ are determined by requiring that the energy-momentum tensor shall be the source and that the right Newtonian limit obtains. The equation must therefore be something like

$$'\Box' g_{\mu\nu} + ? = -\frac{16\pi G}{c^2} \cdot \mathfrak{T}_{\mu\nu} \tag{9.3.1}$$

where $\mathfrak{T}_{\mu\nu}$ is the energy-momentum tensor of everything except gravitation and $'\Box'$ is the equivalent in general coordinates of the d'Alembertian operator in Cartesian coordinates. The ? recognises the probability that the left-hand side will also contain products of the derivatives of $g_{\mu\nu}$, since the gravitational field energy is a source of the gravitational potential. The principle of strong equivalence is then put in by requiring that the equations determining the gravitational field shall themselves be generally covariant, like everything else. There is then only one possible form for the left-hand side of Eq. (9.3.1) which gives the correct Newtonian limit *. Furthermore, in a local freely falling frame where no gravitational effects are apparent, $\mathfrak{T}_{\mu\nu}$ is conserved. This requires the left-hand side of Eq. (9.3.1) to satisfy four identities and this is only possible with one ratio of the two pieces. The set of ten independent equations (9.3.1) (16-6 because $\mathfrak{T}_{\mu\nu}$ is symmetric) with the imposed identities now have four degrees of freedom such that if $g_{\mu\nu}(x)$ is a solution, so is $g'_{\mu\nu}(x')$ where $g'_{\mu\nu}(x')$ may be computed from $g_{\mu\nu}(x)$ and any specified transformation of coordinates $x \rightarrow x'$. Such a transformation relates every x_μ to the corresponding set of x'_λ and of course can never dispose of a permanent gravitational field everywhere: it corresponds to a redefinition of coordinates in a specified curved spacetime. This condition is necessary for general covariance of Eq. (9.3.1), which must, if the principle

---

* A term $\lambda g_{\mu\nu}$ could be added but there is no evidence for the existence of such a term. The quantity λ is known as the cosmological constant.

of strong equivalence is true, be equally valid in any coordinate system. In our flat space theory this condition appeared as gauge invariance. The principle of general covariance applied to the gravitational field equations, strong equivalence, is sufficiently powerful first to require field equations reducing in the weak field limit to the tensor theory (thus predicting the deflection of light by the Sun rather than using it as an input) and secondly to determine without ambiguity the full field equations including the nonlinear effects due to gravity being a source of gravity. The substantial mathematical development necessary to handle all this may be found in many places [2]. It should however be clear that for the solution of a problem in gravitation (for example, planetary motion or the deflection of light by the sun) it is first necessary to choose four conditions which effectively specify the coordinate system in which $g_{\mu\nu}$ is expressed. The motion of a particle or a photon can then be tracked in this coordinate system through Eqs.(9.2.15) and (9.2.7). The coordinate system however has arbitrary elements and in any case is not locally observable. The rest of this chapter is concerned with what the equations of motion in such coordinates actually mean, and how to predict the results of real physical measurements.

9.4 The metric tensor in some simple situations

We want to relate the coordinate system in which $g_{\mu\nu}$ and the equations of motion are expressed to real measurements. We employ for this the invariant interval

$$d\tau^2 = -g_{\mu\nu}dx_\mu dx_\nu \ . \tag{9.4.1}$$

In an infinitesimally small region of spacetime labelled with coordinates x_μ, the laws of physics are the same as in a Lorentz frame of reference momentarily at rest with respect to this coordinate patch. This is how special relativity handles accelerations and it works (see Chapter 1). In particular, a real short measuring rod has the same length in both systems and identical clocks at rest in each system record the same small interval of elapsed time. In the Lorentz frame

$$d\tau^2 = -\eta_{\alpha\beta}d\xi_\alpha d\xi_\beta \ . \tag{9.4.2}$$

A small rod of specified manufacture (for example, of length 10^8 carbon atoms) has length (if at rest)

$$d\ell_\xi^2 = d\xi_i^2 \tag{9.4.3}$$

(i runs from 1 to 3). A clock at rest ticks off

$$dt_\xi^2 = d\xi_0^2 = d\tau^2 \ . \tag{9.4.4}$$

Now

$$dx_\mu = \frac{\partial x_\mu}{\partial \xi_\alpha} d\xi_\alpha$$

or

$$d\xi_a = \frac{\partial \xi_\alpha}{\partial x_\mu} dx_\mu .$$

The quantity $\partial\xi_i/\partial x_0$ is zero because the Lorentz frame is locally at rest with respect to the arbitrary frame. Then

$$d\xi_i = \frac{\partial \xi_i}{\partial x_j} dx_j$$

and the length of a measuring rod at rest is given by

$$(d\xi_i)^2 = \frac{\partial \xi_i}{\partial x_j} dx_j \frac{\partial \xi_i}{\partial x_k} dx_k . \tag{9.4.5}$$

Now

$$g_{\mu\nu} = \eta_{\alpha\beta} \frac{\partial \xi_\alpha}{\partial x_\mu} \frac{\partial \xi_\beta}{\partial x_\nu} = \frac{\partial \xi_i}{\partial x_\mu} \frac{\partial \xi_i}{\partial x_\nu} - \frac{\partial \xi_0}{\partial x_\mu} \frac{\partial \xi_0}{\partial x_\nu}$$

and so

$$g_{oo} = -\left(\frac{\partial \xi_0}{\partial x_0}\right)^2$$

because $\partial\xi_i/\partial x_0$ vanishes, and

$$g_{jk} = \frac{\partial \xi_i}{\partial x_j} \frac{\partial \xi_i}{\partial x_k} - \frac{\partial \xi_0}{\partial x_j} \frac{\partial \xi_0}{\partial x_k} . \tag{9.4.6}$$

Clearly $(d\xi_i)^2 = g_{jk}\, dx_j\, dx_k$ if $\partial\xi_0/\partial x_j$ vanishes. This is not necessarily the case, even though $\partial\xi_i/\partial x_0$ vanishes. We also have

$$g_{ok} = \frac{\partial \xi_i}{\partial x_0} \frac{\partial \xi_i}{\partial x_k} - \frac{\partial \xi_0}{\partial x_0} \frac{\partial \xi_0}{\partial x_k}$$

and so obtain for the case where $\partial\xi_i/\partial x_0$ vanishes

$$\frac{\partial \xi_0}{\partial x_k} = -\frac{g_{ok}}{\sqrt{-g_{oo}}} .$$

Thus the proper length of an infinitesimally short measuring rod [3] is given by

$$d\ell^2 = dx_j\, dx_k \left\{ g_{jk} - \frac{g_{oj} g_{ok}}{g_{oo}} \right\} \tag{9.4.7}$$

and clearly an interval of proper time as recorded by a clock of specified manufacture at rest in the arbitrary frame is given by

$$d\tau^2 = - g_{oo}\, dx_0^2 . \tag{9.4.8}$$

These relations should be general, because we originally defined the $g_{\mu\nu}$ in terms of partial differential coefficients linking a local Lorentz frame in free fall with the specified frame: we did not specify the relative velocity of the frames. We can get the same answer in a different way, [4]. If $dx_i = 0$ then clearly $d\tau^2 = -g_{oo}\, dx_o^2$. For measurement of distance we employ a light pulse transmitted, reflected from a mirror and received at the origin, and time it with a real standard clock. The length between source and mirror is $\Delta\ell$, identical to the length of an identical measure in the comoving inertial frame. The proper length is thus $\Delta\tau/2c$, where $\Delta\tau$ is the time elapsed on either standard clock between transmission of a light pulse and reception of the echo. In the arbitrary coordinates the propagation of light follows the relation

$$g_{\mu\nu}\, dx_\mu\, dx_\nu = 0$$

or

$$g_{oo}\, dx_o^2 + g_{ij}\, dx_i\, dx_j + 2g_{oi}\, dx_o\, dx_i = 0$$

$$dx_o = -\frac{1}{g_{oo}}\left[g_{oi}\, dx_i \pm \sqrt{\left(g_{oi}\, dx_i\, g_{oj}\, dx_j - g_{oo}\, g_{ij}\, dx_i\, dx_j\right)}\right] . \qquad (9.4.9)$$

On the outward journey dx_i is positive, on the return journey negative (or vice versa). The total coordinate time elapsed over this very short distance is thus

$$\Delta x_o = -\frac{2}{g_{oo}}\sqrt{g_{oi}\, dx_i\, g_{oj}\, dx_j - g_{oo}\, g_{ij}\, dx_i\, dx_j} .$$

The corresponding element of proper time recorded by the clock at the source is

$$\Delta\tau = \frac{2}{\sqrt{-g_{oo}}}\sqrt{g_{oi}\, dx_i\, g_{oj}\, dx_j - g_{oo}\, g_{ij}\, dx_i\, dx_j} \qquad (9.4.10)$$

and $\Delta\tau = 2\Delta\ell$ (with $c = 1$) and so

$$\Delta\ell^2 = dx_i\, dx_j \left\{ g_{ij} - \frac{g_{oi}\, g_{oj}}{g_{oo}} \right\} \qquad (9.4.11)$$

once more. It should now be clear that the measured local velocity of light is always the same number c (whether defined as 1 light-sec s^{-1} or as 3×10^{10} cm s^{-1}) even if the coordinate velocity is not. However, if we time light over a distance such that the approximation of constant $g_{\mu\nu}$ is not valid, the measured average velocity of light will <u>not</u> be c .

We will now examine some simple cases in which the arbitrary frame x_μ is related through straight line motion to a specified inertial frame X_λ. First let

$$X = \frac{x - vt}{\sqrt{1 - v^2}} \qquad T = \frac{t - vx}{\sqrt{1 - v^2}} \qquad Y = y \qquad Z = z . \qquad (9.4.12)$$

From these transformations we compute $g_{\mu\nu}$ and find of course $g_{\mu\nu} = \eta_{\mu\nu}$. The coordinate time is the proper time and the space markers are uniformly separated in a well defined proper length. The velocity of light is $c(=1)$ in all directions and over any range, both in terms of the coordinates and as actually measured: the transformation is the Lorentz transformation characterised by

$$\left.\frac{dx}{dt}\right|_X = v .$$

Now suppose

$$X = x - vt \qquad T = t \qquad Y = y \qquad Z = z \tag{9.4.13}$$

Again

$$\left.\frac{dx}{dt}\right|_X = v$$

but this is the Galilean transformation which is abandoned in special relativity. In general relativity it is perfectly admissible. We at once find

$$g_{00} = -(1 - v^2)\ , \quad g_{11} = g_{22} = g_{33} = 1\ , \quad g_{01} = g_{10} = -v \tag{9.4.14}$$

and all other components zero. The derivatives of the metric tensor vanish, so there are no inertial or gravitational forces. The values of the components do not depend on the coordinates and so the velocity of light is equal to c in any direction, measured over any range with a real clock and a real measuring rod. The velocity of light measured in terms of the coordinates does not depend on position, but it is not equal to c and depends on the direction.

$$\left.\frac{dx}{dx_0}\right|_{d\tau^2 = 0} = -\frac{g_{00}}{g_{01} \pm \sqrt{g_{01}g_{01} - g_{00}g_{11}}} = \frac{1 - v^2}{\pm 1 - v} = \pm 1 + v \tag{9.4.15}$$

$$\left.\frac{dy}{dx_0}\right|_{d\tau^2 = 0} = \left.\frac{dz}{dx_0}\right|_{d\tau^2 = 0} = \sqrt{1 - v^2}\ . \tag{9.4.16}$$

The relationship between proper time and coordinate time at any point is

$$d\tau = \sqrt{1 - v^2}\ dt \tag{9.4.17}$$

and

$$\Delta\ell_x = \frac{\Delta x}{\sqrt{1 - v^2}}\ , \quad \Delta\ell_y = \Delta y\ , \quad \Delta\ell_z = \Delta z\ . \tag{9.4.18}$$

It is quite clear what is happening. In both this case and the previous case we are transforming between two inertial frames. In the first case we used the Lorentz transformation which links coordinates measured in standard proper units to coordinates in the second frame measured in standard proper units. In the second case the two reference frames are the same as in the first case, but in the second frame the coordinate clocks are all running at a faster rate than standard clocks so as to allow $T = t$ despite time dilation and the x

coordinates are chosen expanded so as to allow $X = x - vt$ despite the Lorentz contraction. A redefinition of coordinates in this frame takes us back to the first case [5]. This example is helpful in understanding some of the features encountered with accelerated motion.

9.5 An example of an inertial field

Consider a space vehicle with its engines blasting so as to produce a constant acceleration g as measured by an accelerometer on board. The accelerometer is therefore accelerating at g with respect to the comoving inertial frame. We track the vehicle from a given inertial frame by using the four-velocity and four-acceleration. In the local instantaneously comoving inertial frame carrying coordinates ξ_α the proper time coincides with proper time on the vehicle, and so

$$\frac{d^2\xi_1}{d\tau^2} = g \tag{9.5.1}$$

and is the only non-zero component of the four-acceleration. Lorentz transformation of the four-acceleration gives in a specified inertial frame

$$\frac{d^2X}{d\tau^2} = \frac{g}{\sqrt{1-v^2}} \tag{9.5.2}$$

where v is the velocity of the comoving inertial frame in the fixed frame. The four-velocity in the comoving frame has one component

$$\frac{d\xi_0}{d\tau} = 1$$

so

$$\frac{dX}{d\tau} = \frac{v}{\sqrt{1-v^2}} \, . \tag{9.5.3}$$

Then

$$\frac{d}{d\tau}\left(\frac{dX}{d\tau}\right) = \frac{g}{\sqrt{1-v^2}}$$

$$\frac{dT}{d\tau}\,\frac{d}{dT}\left(\frac{v}{\sqrt{1-v^2}}\right) = \frac{g}{\sqrt{1-v^2}}$$

$$d\tau^2 = dT^2 - dX^2$$

so

$$\frac{d\tau}{dT} = \sqrt{1-v^2}$$

whence

$$\frac{d}{dT}\left(\frac{v}{\sqrt{1-v^2}}\right) = g \tag{9.5.4}$$

where

$$v = \left.\frac{dX}{dT}\right|_{\xi_1} .$$

This is the equation of motion of the vehicle in the X frame, and is also the equation of motion of a particle acted on by a constant force, for example an electron in a constant electric field.

It will be more convenient to note that since the magnitude of the four-acceleration is an invariant

$$\left(\frac{d^2X}{d\tau^2}\right)^2 - \left(\frac{d^2T}{d\tau^2}\right)^2 = g^2. \tag{9.5.5}$$

With

$$\left(\frac{dX}{d\tau}\right)^2 - \left(\frac{dT}{d\tau}\right)^2 = -1 \tag{9.5.6}$$

$$\frac{d^2X}{dT^2}\frac{dX}{d\tau} - \frac{d^2T}{d\tau^2}\frac{dT}{d\tau} = 0$$

we find

$$\frac{d^2X}{d\tau^2} = g\frac{dT}{d\tau} \qquad \frac{d^2T}{d\tau^2} = g\frac{dX}{d\tau} \quad . \tag{9.5.7}$$

Differentiate once, substitute and solve. The solutions are exponentials which can be combined to form the hyperbolic functions. Since $T \to \tau$ as $g\tau \to 0$ and $X \to \frac{1}{2}gT^2$, the solutions are

$$X = g^{-1}(\cosh g\tau - 1)\ ,\ T = g^{-1}\sinh g\tau\ ,\quad v = \tanh g\tau\ . \tag{9.5.8}$$

(results which were already implied by the form of (9.5.6)). The quantity τ is proper time as registered on a clock at the same place as the accelerometer reading g. We now equip the astronaut in the vehicle with a framework of rods and clocks reading a coordinate time $\xi_0 = \tau(0)$, which is the proper time in the frame instantaneously comoving with the observer. In this comoving frame we have coordinates ξ_0, ξ_1 (and ξ_2, ξ_3) and freezing the picture instantaneously

$$\left.\begin{aligned} X &= \frac{\xi_1 + v\,\xi_0}{\sqrt{1-v^2}} \\ T &= \frac{\xi_0 + v\,\xi_1}{\sqrt{1-v^2}} \end{aligned}\right\} \tag{9.5.9}$$

The astronaut has coordinates $X(0)$, $T(0)$; $\xi_1(0) = 0$, ξ_0 and the values of X, T corresponding to arbitrary values of ξ_1 are obtained from Eqs. (9.5.8) and (9.5.9):

$$X - X(0) = \xi_1 \cosh g\,\tau(0) = \xi_1 \cosh g\,\xi_0$$

$$T - T(0) = \xi_1 \sinh g\,\tau(0) = \xi_1 \sinh g\,\xi_0$$

at the same instant of coordinate time, whence

$$X = (g^{-1} + \xi_1)\cosh g\,\xi_0 - g^{-1}$$
$$T = (g^{-1} + \xi_1)\sinh g\,\xi_0 \quad . \tag{9.5.10}$$

All we have done is to use special relativity to give our astronaut a reference frame coinciding with that of the comoving inertial frame with Lorentz coordinates: it is for this reason that we keep the labels ξ_α. (If the acceleration is not constant establishing the equivalent frame is much more complicated.) We have from Eq. (9.5.10)

$$dT^2 - dX^2 = (1 + g\,\xi_1)^2\, d\xi_0^2 - d\xi_1^2 \ . \tag{9.5.11}$$

The only term in $g_{\mu\nu}$ different from the corresponding term in $\eta_{\mu\nu}$ is $g_{00} = -(1 + g\,\xi_1)^2$. Thus proper time elapsed depends on position in this very natural coordinate frame, and the acceleration as measured at any given ξ_1 departs from the value g by a factor $(1 + g\,\xi_1)$.

Let us work out accelerations as a function of position. To do this we find the equations of motion of a freely falling particle, using Eq. (9.2.11) and (9.2.14). The only non-zero derivative of the metric tensor $g_{\mu\nu}$ is

$$\frac{\partial g_{00}}{\partial \xi_1} = -\,2g\,(1 + g\,\xi_1)\ . \tag{9.5.12}$$

If we eliminate the proper time on the falling particle from Eq. (9.2.11) we obtain

$$\frac{d^2\xi_i}{d\xi_0^2} = -\Gamma^i_{\nu\mu}\,\frac{d\xi_\nu}{d\xi_0}\,\frac{d\xi_\lambda}{d\xi_0} + \Gamma^0_{\nu\mu}\,\frac{d\xi_\nu}{d\xi_0}\,\frac{d\xi_\lambda}{d\xi_0}\,\frac{d\xi_i}{d\xi_0} \quad . \tag{9.5.13}$$

The only non-zero components of the affine connection are

$$\Gamma^1_{00} = -\tfrac{1}{2}\,\frac{\partial g_{00}}{\partial \xi_1}\, g^{-1}_{11} = g(1 + g\,\xi_1)$$

$$\Gamma^0_{10} = \Gamma^0_{01} = \tfrac{1}{2}\,\frac{\partial g_{00}}{\partial \xi_1}\, g^{-1}_{oo} = g(1 + g\,\xi_1)^{-1}$$

so that

$$\frac{d^2\xi_1}{d\xi_0^2} = -\,g(1 + g\,\xi_1) + 2g(1 + g\,\xi_1)^{-1}\left(\frac{d\xi_1}{d\xi_0}\right)^2 \tag{9.5.14}$$

is the equation of motion of a freely falling particle. Note particularly that in this inertial field the acceleration is independent of the transverse velocity.

The acceleration recorded by an accelerometer at ξ_1 will correspond to $(d\xi_1/d\xi_0) = 0$ but the coordinate time interval $d\xi_0$ is not the local frame proper time interval $d\tau$ except at $\xi_1 = 0$. At fixed ξ_1 we have

$d\tau^2 = (1+g\,\xi_1)^2\,d\xi_0^2$ and so for fixed ξ_1 the acceleration recorded is

$$\frac{d^2\xi_1}{d\xi_0^2}\left(\frac{d\xi_0}{d\tau}\right)^2 = \frac{1}{(1+g\,\xi_1)^2}\,\frac{d^2\xi_1}{d\xi_0^2} = -\,g(1+g\,\xi_1)^{-1} \qquad (9.5.15)$$

for

$$\frac{d\xi_1}{d\xi_0} = 0\,.$$

The space coordinates are uniform and isotropic when measured with real measuring rods: the reason why the acceleration differs from g for $\xi_1 \neq 0$ is that seen from a given inertial frame this coordinate system on the spacecraft is shrinking longitudinally as the spacecraft accelerates — a progressive Lorentz contraction. If the acceleration was g for all ξ_1, as it could be for ξ_1 markers attached to separate rockets firing in an identical pattern, then a given length $\Delta\xi_1$ would increase with time as measured with standard measuring rods.

For small ξ_1, $g\,\xi_0$

$$\begin{aligned} X &\simeq \xi_1 + (\tfrac{1}{2}g\,\xi_0)\,\xi_0 \\ T &= \xi_0 + (\tfrac{1}{2}g\,\xi_0)\,\xi_1 \end{aligned} \qquad . \qquad (9.5.16)$$

This is obviously the equivalent of setting

$$\begin{aligned} X &= \xi_1 + v\,\xi_0 \\ T &= \xi_0 + v\,\xi_1 \end{aligned} \qquad (9.5.17)$$

in the Lorentz transformation (neglecting v^2) rather than using the Galilean transformation

$$\begin{aligned} X &= \xi_1 + v\,\xi_0^2 \\ T &= \xi_0 \end{aligned} \qquad . \qquad (9.5.18)$$

It is for this reason that g_{01} vanishes: if we assumed (for small $g\,\xi_0$) a transformation

$$\begin{aligned} X &= \xi_1 + \tfrac{1}{2}g\,\xi_0^2 \\ T &= \xi_0 \end{aligned}$$

then we would have

$$g_{11} = 1\,, \quad g_{00} = -1(1-g^2\xi_0^2)\,, \quad g_{01} = g\,\xi_0\,.$$

The proper acceleration, as measured by the acceleration in the ξ frame of a test particle released from rest, is the same in both cases.

The coordinate system constructed from special relativity has the advantage that the inertial force appears as the gradient of a time independent potential, thus naturally having this feature in common with a force due to a permanent gravitational field.

Let us confine ourselves to a region sufficiently close to an accelerometer reading g that we may set

$$g_{00} = -(1 + 2g\,\xi_1)\ .$$

The coordinate velocity of light is thus given by

$$(d\underline{\xi})^2 - (1 + 2g\,\xi_1)\,d\xi_0^2 = 0\ , \qquad \frac{d\xi_1}{d\xi_0} = 1 + g\,\xi_1$$

where $\underline{\xi}$ is measured from the accelerometer. Because $d\tau = (1 + g\,\xi_1)\,d\xi_0$ at a given ξ_1, the round trip time over a very short path clearly gives a measured local velocity $c = 1$ everywhere.

We can calculate the bending of a ray of light in the ξ coordinates using the coordinate velocity. If we consider a plane wave initially propagating at right angles to the acceleration, say along ξ_2, then the arguments of section 5.1 yield

$$\frac{d^2\xi_1}{d\xi_2^2} = -\frac{1}{c}\frac{\partial c}{\partial \xi_1} \tag{9.5.19}$$

and the angle

$$\frac{d\xi_1}{d\xi_2} = -g\,\Delta\,\xi_2$$

after traversing $\Delta\,\xi_2$.

This is in agreement with the result of section 1.2. Note that this is the angular deflection that would actually be measured locally because the space coordinates ξ_1 are uniformly spaced as measured by standard measuring rods.

From g_{00} we may also compute the redshift. The quantity ξ_0 is a universal coordinate time. Consider two standard clocks at rest with respect to the coordinates $\underline{\xi}$, one at $\xi_1 = 0$ and the other at a particular value of ξ_1. Corresponding intervals of proper time are given by

$$d\tau(\xi_1) = (1 + g\,\xi_1)\,d\xi_0$$

$$d\tau(0) = d\xi_0\ .$$

Let the proper frequency of an atom be ν. Let $\Delta\tau(0)$ be the time between two successive pulses of such an atom at rest at $\xi = 0$. Because in these coordinates $g_{\mu\nu}$ is independent of ξ_0 and the atom is at rest, these two pulses of radiation are spaced by $\Delta\,\xi_0 = \Delta\tau(0)$ at ξ_1. In this same interval of coordinate time the proper time interval at ξ_1 is $\Delta\tau(\xi_1) = (1 + g\,\xi_1)\,\Delta\xi(0)$

in which time $(1+g\,\xi_1)$ pulses are emitted by an identical atom at rest at ξ_1. Then

$$\frac{\nu(\xi_1)}{\nu(0)} = 1 + g\,\xi_1 \qquad (9.5.20)$$

which again agrees with the results of section 1.2. Note that g can be measured as the acceleration at $\xi_1 = 0$ and ξ_1 is a distance that can be directly measured with a real measuring stick.

This section has been concerned with the inertial field appearing in an accelerated laboratory. In a permanent gravitational field, additional features are present because the spatial part of the metric tensor is no longer δ_{ij}.

9.6 The external spherically symmetric gravitational field

The Einstein field equations relate the metric tensor $g_{\mu\nu}$ to the energy-momentum tensor $\mathfrak{J}_{\mu\nu}$ and determine the quantities $g_{\mu\nu}$ up to a coordinate transformation. The quantity on the left-hand side of Eq. (9.3.1) contains second derivatives of $g_{\mu\nu}$ corresponding to $\Box\, h_{\mu\nu}$, and bilinear products of first derivatives, corresponding to the energy-momentum tensor of the gravitational field itself. $\mathfrak{J}_{\mu\nu}$ is thus the energy-momentum tensor of everything except gravity: solution of the Einstein field equations in a region where $\mathfrak{J}_{\mu\nu} = 0$ corresponds to the solution of the Newtonian equation in empty space,

$$\nabla^2\varphi = 0 .$$

The latter equation is trivial to solve for the case of spherical symmetry yielding

$$\varphi = \frac{K}{r}$$

and we identify the constant K with $-GM$ where M is the (spherically symmetric) mass at the origin.

The corresponding solution of the Einstein field equations is obtained as follows [6]. First, set $\mathfrak{J}_{\mu\nu} = 0$. Secondly, choose a coordinate system. If the coordinate system is taken to be in the rest frame of the source, it seems natural to choose coordinates such that $g_{\mu\nu}$ does not depend on time. We would also like time-orthogonal coordinates: that is, g_{oi} vanishes. For a spherically symmetric field it is attractive to leave the tangential parts of a small interval alone, and write

$$d\tau^2 = B(r)\,dt^2 - A(r)\,dr^2 - r^2(d\theta^2 + \sin^2\theta\,d\varphi^2) \qquad (9.6.1)$$

in spherical polar coordinates, or

$$g_{rr} = A(r)\,, \quad g_{\theta\theta} = r^2\,, \quad g_{\varphi\varphi} = r^2\sin^2\theta\,, \quad g_{oo} = -B(r)\,.$$

The effect of the gravitational field is embodied in g_{rr} and g_{oo}.

This choice of coordinates satisfies the Einstein field equations with $\mathfrak{J}_{\mu\nu} = 0$ provided that

$$\left.\begin{aligned} g_{oo} &= -\left[1 + \frac{K}{r}\right] \\ g_{rr} &= \left[1 + \frac{K}{r}\right]^{-1} \end{aligned}\right\} . \qquad (9.6.2)$$

Since the gravitational field is given in the Newtonian limit by $\frac{1}{2}\nabla g_{oo}$, the constant K is identified with $-2MG$, where M is the mass contained within r as $r \to \infty$. This is the Schwarzschild exterior solution in standard coordinates

$$d\tau^2 = \left(1 - \frac{2GM}{r}\right) dt^2 - \frac{dr^2}{1 - \frac{2GM}{r}} - r^2\, d\theta^2 - r^2 \sin^2\theta\, d\varphi^2 \ . \qquad (9.6.3)$$

Time is affected, radial lengths are affected but tangential lengths are not. In Chapter 6 we found that all short lengths were equally affected by a gravitational field. It is therefore not surprising to find that we can redefine the radial coordinate r so as to achieve this as a property of the Schwarzschild solution. Set

$$r = \rho \left(1 + \frac{MG}{2\rho}\right)^2 \qquad (9.6.4)$$

when

$$1 - \frac{2GM}{r} = \left(\frac{1 - \frac{MG}{2\rho}}{1 + \frac{MG}{2\rho}}\right)^2$$

$$dr = d\rho\left(1 + \frac{MG}{2\rho}\right)\left(1 - \frac{MG}{2\rho}\right)$$

so that

$$d\tau^2 = \left(\frac{1 - \frac{MG}{2\rho}}{1 + \frac{MG}{2\rho}}\right)^2 dt^2 - \left(1 + \frac{MG}{2\rho}\right)^4 \left(d\rho^2 + \rho^2\, d\theta^2 + \rho^2 \sin^2\theta\, d\varphi^2\right) \qquad (9.6.5)$$

This expresses the Schwarzschild solution in isotropic coordinates. If the Schwarzschild metric is used in the standard form, the equations of motion resulting are closely analogous to the Newtonian equations of motion. If isotropic coordinates are used measurements over small distances with real rods are related to the coordinate intervals merely by an isotropic change of scale. Furthermore in the weak field limit we obtain precisely the tensor theory of gravity developed in the first part of this book from the equations

$$\Box\, h_{\mu\nu} = 8\pi G \bar{\mathfrak{J}}_{\mu\nu}$$

$$\frac{\partial \mathfrak{J}_{\mu\nu}}{\partial x_\mu} = 0 \ .$$

Both systems of coordinates are widely used and neither is in any sense the more fundamental. There is a third system of coordinates also frequently encountered, the so-called harmonic system of coordinates [7] in which

$$\left.\begin{aligned} g_{oo} &= -\frac{1-\frac{MG}{R}}{1+\frac{MG}{R}} \qquad g_{io} = 0 \\ g_{ij} &= \left(1+\frac{MG}{R}\right)^2 \delta_{ij} + \left(\frac{MG}{R}\right)^2 \frac{1+\frac{MG}{R}}{1-\frac{MG}{R}} \left(\frac{x_i x_j}{R^2}\right) \end{aligned}\right\} . \tag{9.6.6}$$

To first order in the usually small quantity GM/R or GM/ρ, g_{ij} and g_{oo} in the isotropic coordinate system and the harmonic coordinate system agree. The forms of g_{oo} also agree to second order. Consequently the two systems are equivalent for the solar system measurements we have considered and we shall discuss the harmonic system no further. However, the standard and isotropic forms of the metric differ in first order in $g_{\theta\theta}$ and $g_{\varphi\varphi}$ and in second order in g_{oo}. The physical content is nonetheless the same. We shall work out the predictions for the measurable quantities already discussed, namely the gravitational redshift, deflection of light by the Sun, radar echo delay and the precession of the perihelion of Mercury, using both coordinate systems.

9.7 The gravitational redshift

Consider first the gravitational redshift. An interval of proper time is related to an interval of coordinate time through

$$d\tau = \sqrt{-g_{oo}}\, dt \tag{9.7.1}$$

for a clock at rest in the coordinate system. At two different levels in the gravitational potentials different intervals of proper time correspond to the same interval of coordinate time. Let two pulses of light be emitted from a source at a radial coordinate r_1, an interval of coordinate time Δt apart. The elements of the metric tensor are independent of coordinate time so that the coordinate time separation at a radial coordinate r_2 is still Δt.

$$\Delta t = \frac{\Delta\tau(r_1)}{\sqrt{-g_{oo}(r_1)}} \quad , \quad \Delta\tau(r_2) = \sqrt{-g_{oo}(r_2)}\, \Delta t$$

so

$$\frac{\Delta\tau(r_2)}{\Delta\tau(r_1)} = \sqrt{\frac{g_{oo}(r_2)}{g_{oo}(r_1)}} . \tag{9.7.2}$$

To first order in GM/r, g_{00} has the same form for either isotropic or standard coordinates: using the standard form

$$\frac{\Delta\tau(r_2)}{\Delta\tau(r_1)} = \sqrt{\frac{1 - \frac{2GM}{r_2}}{1 - \frac{2GM}{r_1}}} \; . \tag{9.7.3}$$

The frequency of signals ν_2 received at r_2, measured with a real standard clock, is related to the frequency of the same signals emitted at r_1, measured with a real standard clock, by

$$\frac{\nu_2}{\nu_1} = \sqrt{\frac{1 - \frac{2GM}{r_1}}{1 - \frac{2GM}{r_2}}} \simeq 1 + GM\left\{\frac{1}{r_2} - \frac{1}{r_1}\right\}$$

or

$$\frac{\nu_2 - \nu_1}{\nu} \simeq GM\left\{\frac{1}{r_2} - \frac{1}{r_1}\right\} \; . \tag{9.7.4}$$

The frequencies ν_1 and ν_2 are measurable quantities. The right-hand side of this equation however contains the coordinates r_2 and r_1 and a mass M. In order to predict the redshift we need to reduce the right-hand side to measured quantities. Over small distances

$$\frac{\nu_2 - \nu_1}{\nu} \simeq -\frac{GM}{r^2}\,\Delta r \qquad \Delta r = r_2 - r_1 \; .$$

The quantity Δr is a coordinate distance. The radial distance ΔL between r_1 and r_2 measured with a real measuring rod is related to Δr by

$$\Delta L = \sqrt{g_{rr}}\;\Delta r \simeq \Delta r \tag{9.7.5}$$

if we only want the expression for the redshift to first order. We may therefore replace the coordinate interval Δr by ΔL, measured with a real measuring stick. Finally, the acceleration of a freely falling particle is given by Eq. (9.2.11) and for negligible velocity we have

$$\frac{d^2 r}{dt^2} = -\frac{GM}{r^2} \; . \tag{9.7.6}$$

The measured acceleration g is equal to d^2r/dt^2 if we ignore corrections of order GM/r and higher powers, so the redshift recipe is, to first order in the gravitational potential,

$$\frac{\Delta\nu}{\nu} \simeq g\,\Delta L \tag{9.7.7}$$

an expression which contains only measured quantities. This frequency shift clearly disappears if the laboratory containing the source and the detector is in free fall instead of fixed in the coordinates.

9.8 Deflection of light by the Sun

We will calculate the deflection of light by setting $d\tau^2 = 0$ for a pulse or wavefront moving at the velocity of light. The coordinate velocity of light is then given by

$$g_{\mu\nu} \, dx_\mu \, dx_\nu = 0 \quad . \tag{9.8.1}$$

The local velocity of light, as measured by timing over an infinitesimal distance, using standard clocks and rods, is always c .

Consider a plane wave propagating at right angles to the gravitational field in a small laboratory. Label the radial coordinate by y and the tangential coordinate by x . Then the angle turned through in distance Δx is given by

$$\frac{dy}{dx} = -\frac{1}{u}\frac{\partial u}{\partial y}\,\Delta x \tag{9.8.2}$$

where u is the tangential coordinate velocity of light.

In isotropic coordinates the coordinate velocity of light in any direction is

$$\left(1-\frac{MG}{2r}\right)\left(1+\frac{MG}{2r}\right)^{-3} \simeq 1-\frac{2GM}{r} \tag{9.8.3}$$

so

$$\frac{dy}{dx} = \frac{2GM}{y^2}\,\Delta x = \frac{2GM}{r^2}\,\Delta x \quad . \tag{9.8.4}$$

Because the coordinates are isotropic, this angle preserves its value in real units. But we measure it with a telescope clamped to the far wall of the laboratory of dimension Δx . Now

$$\Delta L_x = \left(1+\frac{GM}{2r}\right)^2 \Delta x \sim \left(1+\frac{GM}{r}\right)\Delta x \quad . \tag{9.8.5}$$

So the walls of the laboratory are set at a small angle to the coordinate frame, just enough for the telescope to measure an angle

$$\frac{GM}{r^2}\,\Delta L_x \qquad \text{(compare section 6.2)}$$

where ΔL_x is the measured distance across the laboratory, and the quantity $(GM/r^2) = g$ is the measured acceleration of a slowly moving particle. (This result is again only to first order in GM/r.) The variation of $\left(1+\frac{GM}{2r}\right)^2$ in relating the measurement system to the coordinates can only be ignored if the first derivative of GM/r can be ignored — in which case to this approximation there is no deflection.

In standard coordinates the picture is a little different. The tangential velocity is

$$u = \sqrt{1-\frac{2GM}{r}} \sim 1-\frac{GM}{r} \tag{9.8.6}$$

so

$$\frac{dy}{dx} = \frac{GM}{r^2}\Delta x \ . \qquad (9.8.7)$$

A tangential length $\Delta L_x = \Delta x$ because $g_{\theta\theta}$ and $g_{\varphi\varphi}$ have their flat space values. A length ΔL_y is related to Δy by

$$\Delta L_y = \frac{\Delta y}{\sqrt{1 - \frac{2GM}{r}}}$$

so

$$\frac{dy}{dx} = \frac{GM}{r^2}\Delta L_x$$

if we only want an answer to first order. The side walls of the laboratory are parallel to the y coordinate in this case and the telescope measures an angle $g\,\Delta L_x$ once more.

Thus if we calculate the result of a measurement of the deflection of light in a small laboratory at rest with respect to the source of mass M, we get an answer independent of which set of coordinates we used, and the answer satisfies the principle of equivalence.

The coordinate deflection of light starting its journey very far from the deflecting mass and being received very far from the deflecting mass can be calculated by integrating the deflection over an (approximately) straight path, as in section 1.4. In isotropic coordinates the velocity is, to first order, $1 - 2\frac{GM}{r}$. It is clear from comparison with section 5.1 that the deflection in isotropic coordinates is then given by $\alpha = \frac{4GM}{b}$ where b is the impact parameter, approximately equal to the distance of closest approach.

In standard coordinates the coordinate velocity of light is direction dependent and the treatment is a little more complicated. We can either solve the equations of motion (9.2.15) or simply use the variation of the coordinate velocity of light. In the latter approach we take the condition for light

$$d\tau^2 = 0 = \left(1 - \frac{2GM}{r}\right) dt^2 - \frac{dr^2}{1 - \frac{2GM}{r}} - r^2\left(d\theta^2 + \sin^2\theta\, d\varphi^2\right) \qquad (9.8.8)$$

and set

$$\frac{d\alpha}{dx} = \frac{1}{u}\frac{du}{dy} \simeq \frac{\partial}{\partial y}\left(\frac{dx}{dt}\right)$$

as before. Here x is the distance along the path, with $\frac{dx}{dt} \approx 1$. We choose $\theta = 90^\circ$, so $d\theta = 0$, and have

$$0 = dt^2\left(1 - \frac{2GM}{r}\right) - r^2\, d\varphi^2 - \frac{dr^2}{1 - \frac{2GM}{r}}$$

which we write as

$$0 = \frac{dx^2 \left(\frac{dr}{dx}\right)^2}{1 - \frac{2GM}{r}} + r^2 \left(\frac{d\varphi}{dx}\right)^2 dx^2 - \left(1 - \frac{2GM}{r}\right) dt^2 \qquad (9.8.9)$$

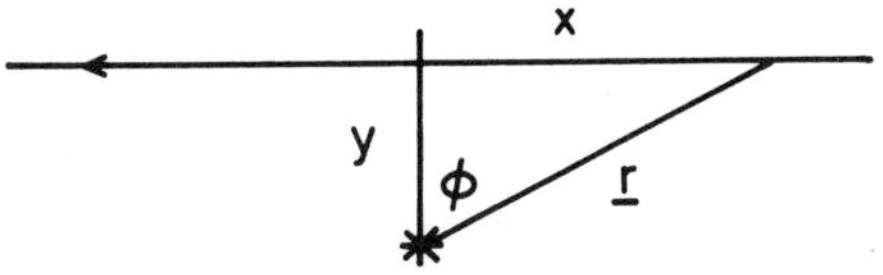

Fig.9.8.1 Coordinates employed in calculating the deflection of light by the Sun, Eq. (9.8.9)

(see Fig.9.8.1) with

$$r = \sqrt{x^2 + y^2}, \quad \frac{dr}{dx} = \frac{x}{r}, \quad \cos\varphi = \frac{y}{r}, \quad \frac{d\varphi}{dx} = \frac{y}{r^2}.$$

We now have

$$0 \simeq dx^2 \left(1 + \frac{2\,GM\,x^2}{r^3}\right) - \left(1 - \frac{2GM}{r}\right) dt^2 \qquad (9.8.10)$$

(to first order) and the velocity of light along its nearly straight trajectory is

$$\frac{dx}{dt} \simeq 1 - \frac{GM}{r}\left(1 + \frac{x^2}{r^2}\right) . \qquad (9.8.11)$$

When $x = 0$ this is the tangential velocity, $1 - \frac{GM}{r}$, and when $x \approx r$ it is the radial velocity $1 - \frac{2GM}{r}$.

Then

$$\alpha = \int_{-\infty}^{\infty} \frac{GM}{r}\left(\frac{3\,x^2\,y}{r^4} + \frac{y}{r^2}\right) dx = \frac{4GM}{y} \rightarrow \frac{4GM}{b} \qquad (9.8.12)$$

once more. The change in angle with b relative to a fixed direction (for example a source not significantly affected) is thus given in either standard or isotropic coordinates by $\frac{4GM}{b}$, to first order in $\frac{GM}{b}$. The angular change is identical to that actually measured in isotropic coordinates: in standard coordinates the difference in the coordinate angle and the measured angle is $0\left(\frac{GM}{R}\right)$ where R is the radius of the Earth's orbit: such corrections, of the order of one part in 10^8, can be ignored in this problem.

9.9 Radar echo delay

The excess time delay encountered in radar echo experiments is more instructive. The time for a round trip between the Earth and Mercury is

$$t = 2 \int_{r_e}^{r_m} \left(\frac{dx}{dt}\right)^{-1} dx . \qquad (9.9.1)$$

In isotropic coordinates we set

$$\frac{dx}{dt} = 1 - \frac{2GM}{r}$$

as in section 5.2 and obtain

$$t_I = 2(x_e + x_m) + 4\,GM_{\odot}\,\ln\left[\frac{4r_e\,r_m}{b^2}\right] \qquad (9.9.2)$$

where b is defined in Fig.5.2.1 and is approximately the impact parameter.

In standard coordinates, using Eq. (9.7.11) we have

$$t_S = 2(x_e + x_m) = 2\int_{r_e}^{r_m} \frac{GM_\odot}{r}\left\{1 + \frac{x^2}{r^2}\right\} dx$$

$$t_S = 2(x_e + x_m) + 4GM_\odot \ln\left[\frac{4r_e r_m}{b^2}\right] - 4GM_\odot . \qquad (9.9.3)$$

In these expressions we have set x_e, x_m equal to r_e and r_m in the logarithmic terms. We can express x_e and x_m in the zero order term through $x^2 = \sqrt{r^2 - b^2}$ but it is better to express it in terms of the distance of closest approach to the Sun, r_0. Because the angle turned through in going from $x = 0$ to $x \gg r_0$ is approximately $2GM/r_0$, we have

$$r_0 = \left(\frac{2GM}{r_0} + \beta\right) x_m = \left(\frac{2GM}{r_0} - \beta\right) x_e$$

where β is non-zero because $r_e \neq r_m$. Then

$$x_e = \sqrt{r_e^2 - b^2} = \sqrt{r_e^2 - \left(r_0 + (b - r_0)\right)^2}$$

and

$$x_e + x_m \simeq \sqrt{r_e^2 - r_0^2} + \sqrt{r_m^2 - r_0^2} + 4\,GM$$

and

$$t_I = 2\left(\sqrt{r_e^2 - r_0^2} + \sqrt{r_m^2 - r_0^2}\right) + 4GM_\odot \ln\left(\frac{4r_e r_m}{r_0^2}\right) + 8\,GM_\odot \qquad (9.9.4)$$

$$t_S = 2\left(\sqrt{r_e^2 - r_0^2} + \sqrt{r_m^2 - r_0^2}\right) + 4GM_\odot \ln\left(\frac{4r_e r_m}{r_0^2}\right) + 4\,GM_\odot . \qquad (9.9.5)$$

These two answers are different, but are at once reconciled by remembering that isotropic coordinates are constructed from standard coordinates by the transformation

$$r_S = r_I\left(1 + \frac{MG}{2r_I}\right)^2 \simeq r_I + GM \qquad (9.9.6)$$

leaving everything else alone.

The quantities t, r_e, r_m and r_0 are not directly measurable: rather they are convenient parameters related to directly measurable quantities. The time interval between the transmission anu reception of a radar pulse is measured with an earth-based atomic clock which is running at a different rate from coordinate time. There are two ingredients in the transformation. Because the clock is in the Sun's potential (and indeed in the potential of the Earth) we pick up a factor $\sim 1 - (GM_\odot/r_e)$ and because of the orbital velocity of the Earth relative to the coordinates we pick up a factor $\sqrt{1 - (v^2/c^2)}$, where

$$\frac{v^2}{c^2} = \frac{GM_\odot}{r_e} .$$

The measured proper time interval t_p is thus related to the coordinate time interval t_c by

$$t_p = t_c\left(1 - \tfrac{3}{2}\frac{GM_\odot}{r_e}\right) \qquad (9.9.7)$$

(to first order in $GM_\odot$).

The total travel time is $\sim 1.2 \times 10^3$ secs for the Earth-Mercury-Earth trip and the correction factor is $\sim 1.5 \times 10^{-8}$, introducing a difference between t_p and $t_c \sim 20\,\mu s$, which is not negligible since the total excess delay due to gravitation is $\sim 200\,\mu s$.

The parameters r_e and r_m are not directly measured either: they must be related to something which can be measured. For example, measurement of the time delay when Mercury is at inferior conjunction (between the Earth and the Sun) and at extreme elongation provides two times in terms of which r_e and r_m can be evaluated, using of course the equation of motion of light in the Schwarzschild field. More generally, observations of the time delay can be made over a period of one or more years and the equations of motion of light tested against these results, with r_e and r_m as parameters to be fitted along with many others. Alternatively we may use the directly measurable orbital periods and the equations of motion of the planets (expressed in Schwarzschild coordinates) to determine r_e and r_m. In isotropic coordinates the equations of motion are those we used for calculating the precession of the perihelion of Mercury in Chapter 7 so that the period of revolution

$$T_\theta = 2\pi\sqrt{\frac{R^3}{GM}}\left\{1 - \left(\tfrac{1}{2} - \alpha\right)\frac{GM}{R}\right\} \qquad (9.9.8)$$

where $\alpha = 2$. The substitution $R = R_0 - GM$ takes us to standard coordinates when

$$T_\theta = 2\pi\sqrt{\frac{R_0^3}{GM}} \qquad (9.8.9)$$

which is the Newtonian expression. We thus see (indirectly) that the coordinate time period is related to the coordinate radius of the orbit by the expression obtaining in Newtonian gravitation, if we use standard coordinates rather than isotropic. Again it must be remembered that T_θ is coordinate time and the period measured with an earth-based atomic clock must be corrected to give T_θ. The quantity $GM_\odot$ represents a distance of 1.5 km and a time of travel at the velocity of light of $5\,\mu s$.

It should be clear that the problem of testing the predictions of gravitation for radar echo delay is perfectly well defined in terms of measurable quantities, but of great technical (rather than conceptual) complexity. Interesting detailed discussions exist [8]: we only remark that the predictions of Einstein's theory are currently verified at about the 3% level [9].

9.10 The precession of planetary perihelia

The precession of planetary perihelia is more complicated to calculate, since we need the equation of motion to second order in small quantities. However, comparison of the equations of motion and the Schwarzschild metric tensor in isotropic coordinates with the weak field tensor equations of Chapters 6 and 7 (with the non-linear parameter $\alpha = 2$) shows that we have already solved the problem in isotropic coordinates: the advance is $6\pi \frac{GM}{R}$ radians per revolution.

In standard coordinates we write

$$d\tau^2 = \left(1 - \frac{2GM}{r}\right) dt^2 - r^2 \sin^2\theta \, d\varphi^2 - r^2 \, d\theta^2 - dr^2 \left(1 - \frac{2GM}{r}\right)^{-1} = - g_{\mu\nu} dx_\mu dx_\nu \tag{9.10.1}$$

so that

$$\left.\begin{aligned} g_{tt} &= -\left(1 - \frac{2GM}{r}\right) & g_{\varphi\varphi} &= r^2 \sin^2\theta \\ g_{\theta\theta} &= r^2 & g_{rr} &= \left(1 - \frac{2GM}{r}\right)^{-1} \end{aligned}\right\} . \tag{9.10.2}$$

It should be clear that since g_{rr} is a potential which couples to the square of velocity, we need g_{rr} only to first order in GM/r, but g_{tt} to second order. In contrast with isotropic coordinates however, there is no nonlinear term in g_{tt} in standard coordinates. We shall obtain the equations of motion in standard coordinates and then evaluate the precession rate by the approximate methods of Chapter 7. We work in polar coordinates, both because polar coordinates are appropriate to the problem and because $g_{\mu\nu}$ is diagonal in polar coordinates: the standard form is not diagonal when expressed in rectangular coordinates.

In principle we must compute the affine fields and then solve

$$\frac{d^2 x_\lambda}{d\tau^2} + \Gamma^\lambda_{\mu\nu} \frac{dx_\mu}{d\tau} \frac{dx_\nu}{d\tau} \qquad (9.2.11) \tag{9.10.3}$$

but we can employ some short cuts. We choose $\theta = \frac{\pi}{2}$ to define the plane of the orbit and so are not interested in the equation for θ. Further, if we find the quantities $dt/d\tau$ and $d\varphi/d\tau$ we can then use the expression for proper time to give us the equation of motion for r. We therefore only need to calculate the nonvanishing members of $\Gamma^t_{\mu\nu}$ and $\Gamma^\varphi_{\mu\nu}$.

Now

$$\Gamma^\sigma_{\lambda\mu} = \tfrac{1}{2} g^{\nu\sigma} \left\{ \frac{\partial g_{\mu\nu}}{\partial x_\lambda} + \frac{\partial g_{\lambda\mu}}{\partial x_\mu} - \frac{\partial g_{\mu\lambda}}{\partial x_\nu} \right\} \qquad (9.2.14) \tag{9.10.4}$$

and for diagonal $g_{\nu\sigma}$ the elements of the inverse matrix $g^{\nu\sigma}$ are just the reciprocals of the elements of $g_{\nu\sigma}$. We have

$$\frac{d^2t}{d\tau^2} + \left(\Gamma^t_{rt} + \Gamma^t_{tr}\right)\left(\frac{dt}{d\tau}\right)\left(\frac{dr}{d\tau}\right) = 0 \ , \qquad (9.10.5)$$

all other terms vanishing, with

$$\Gamma^t_{rt} = \Gamma^t_{tr} = \frac{1}{2g_{tt}} \ \frac{\partial}{\partial r} g_{tt} = \frac{1}{1 - \frac{2GM}{r}} \ \frac{GM}{r^2} \qquad (9.10.6)$$

and with θ constant, $d\theta/d\tau$ zero, we have

$$\frac{d^2\varphi}{d\tau^2} + \left(\Gamma^\varphi_{r\varphi} + \Gamma^\varphi_{\varphi r}\right)\left(\frac{d\varphi}{d\tau}\right)\left(\frac{dr}{d\tau}\right) = 0 \ , \qquad (9.10.7)$$

all other terms vanishing, with

$$\Gamma^\varphi_{r\varphi} = \Gamma^\varphi_{\varphi r} = \frac{1}{2g_{\varphi\varphi}} \ \frac{\partial}{\partial r} g_{\varphi\varphi} = \frac{1}{r} \ . \qquad (9.10.8)$$

Then

$$\frac{\frac{d^2t}{d\tau^2}}{\frac{dt}{d\tau}} + \frac{\frac{2GM}{r^2}}{1 - \frac{2GM}{r}} \ \frac{dr}{d\tau} = 0 \qquad (9.10.9)$$

whence

$$\ell n \frac{dt}{d\tau} + \ell n \left(1 - \frac{2GM}{r}\right) = \text{constant} \qquad (9.10.10)$$

and

$$d\tau = K\left(1 - \frac{2GM}{r}\right) dt \qquad (9.10.11)$$

where the value of K is determined by the velocity of the particle. Similarly

$$\frac{\frac{d^2\varphi}{d\tau^2}}{\frac{d\varphi}{d\tau}} + \frac{2}{r} \ \frac{dr}{d\tau} = 0 \qquad (9.10.12)$$

whence

$$\ell n \frac{d\varphi}{d\tau} + 2 \, \ell n \, r = \text{constant} \qquad (9.10.13)$$

or

$$r^2 \frac{d\varphi}{d\tau} = \text{constant} = H \ . \qquad (9.10.14)$$

This equation clearly expresses conservation of angular momentum. With

$$d\tau^2 = \left(1 - \frac{2GM}{r}\right) dt^2 - dr^2 \left(1 - \frac{2GM}{r}\right)^{-1} - r^2 d\varphi^2$$

we can express the radial equation of motion as a function of either t or τ:

$$1 = \frac{\left(1 - \frac{2GM}{r}\right)^{-1}}{K^2} - \left(\frac{dr}{d\tau}\right)^2\left(1 - \frac{2GM}{r}\right)^{-1} - \frac{H^2}{r^2} \qquad (9.10.15)$$

or

$$K^2\left(1 - \frac{2GM}{r}\right)^2 = \left(1 - \frac{2GM}{r}\right) - \left(\frac{dr}{dt}\right)^2\left(1 - \frac{2GM}{r}\right)^{-1} - \frac{H^2}{r^2}\left(1 - \frac{2GM}{r}\right)^2 K^2 \ .$$

For the moment we work with the latter form, which we write as

$$K^2 = \frac{1}{1-\frac{2GM}{r}} - \left(\frac{dr}{dt}\right)^2\left(1-\frac{2GM}{r}\right)^{-3} - \frac{h^2}{r^2} \tag{9.10.16}$$

where

$$h = r^2\frac{d\varphi}{dt}\left(1-\frac{2GM}{r}\right)^{-1}. \tag{9.10.17}$$

Differentiate (9.10.16) with respect to time and obtain

$$0 = -\frac{1}{\left(1-\frac{2GM}{r}\right)^2}\frac{2GM}{r^2}\left(\frac{dr}{dt}\right) - 2\frac{dr^2}{dt^2}\left(\frac{dr}{dt}\right)\left(1-\frac{2GM}{r}\right)^{-3} + 2\frac{h^2}{r^3}\left(\frac{dr}{dt}\right) + 0\left(\left(\frac{dr}{dt}\right)^2\frac{GM}{r}\right). \tag{9.10.18}$$

We drop the last term and write

$$\frac{d^2r}{dt^2} = -\frac{GM}{r^2}\left(1-\frac{2GM}{r}\right) + \frac{h^2}{r^3}\left(1-\frac{2GM}{r}\right)^3. \tag{9.10.19}$$

For a circular orbit $\left.\frac{d^2r}{dt^2}\right|_R = 0$ and so

$$\frac{h^2}{R^3} = \frac{GM}{R^2}\left(1-\frac{2GM}{R}\right)^{-2}. \tag{9.10.20}$$

From Eqs. (9.10.17) and (9.10.20),

$$\frac{d\varphi}{dt} = \sqrt{\frac{GM}{r^3}}$$

and the rotational period in standard coordinates is

$$T_\varphi = 2\pi\sqrt{\frac{R^3}{GM}} \tag{9.10.21}$$

which is the same expression obtaining in Newtonian theory (see Eq. (9.9.9)). We now set $r = \rho + R$ and obtain

$$\frac{d^2\rho}{dt^2} = -\rho\frac{GM}{R^3}\left(1-\frac{6GM}{R}\right) \tag{9.10.22}$$

and the radial period T_R is given by

$$T_R = 2\pi\sqrt{\frac{R^3}{GM}}\left(1+\frac{3GM}{R}\right). \tag{9.10.23}$$

The advance of the perihelion each revolution is then

$$2\pi\left\{\frac{T_R}{T_\varphi} - 1\right\} = 6\pi\frac{GM}{R}$$

once more.

References

[1] S. Weinberg, 'Gravitation and Cosmology', (Wiley 1972), Chapter 3.

[2] For example:
H.A. Atwater, 'Introduction to General Relativity', (Pergamon 1974).
S. Weinberg, 'Gravitation and Cosmology', (Wiley 1972).
C.W. Misner, K.S. Thorne and J.A. Wheeler, 'Gravitation', (Freeman 1973).
C. Møller, 'The Theory of Relativity', (2nd ed., Oxford 1972).
L.D. Landau and E.M. Lifschitz, 'The Classical Theory of Fields', (Pergamon 1962).

[3] C. Møller, 'The Theory of Relativity', (2nd ed., Oxford 1972) Section 8.8.

[4] L.D. Landau and E.M. Lifschitz, 'The Classical Theory of Fields', (Pergamon 1962), p272.

[5] C. Møller, 'The Theory of Relativity', (2nd ed., Oxford 1972) Section 8.14.

[6] See for example any of the standard texts listed under [2].

[7] S. Weinberg, 'Gravitation and Cosmology', (Wiley 1972), p181.

[8] I.I. Shapiro, Phys. Rev., 141, 1219 (1966)
Phys. Rev., 145, 1005 (1966)
D.K. Ross and L.I. Schiff, Phys. Rev., 141, 1215 (1966).

[9] I.I. Shapiro, Phys. Rev. Lett., 26, 1132 (1971).
J.D. Anderson et al., Ap. J., 200, 221 (1975).

CHAPTER 10
BLACK HOLES

10.1 Strong gravitational fields

In the development of the relativistic field theory of gravity, we went only as far as first order in the gravitational potential. We tackled the problem of the precession of planetary perihelia, which involved second order terms, by supposing gravitation to obey the principle of strong equivalence, but without demonstrating that an internally consistent theory results.

The Einstein field equations, in principle, are exact and thus provide us with the means to calculate the effect of such gravitational fields of any strength. We must recognise however that all tests so far made have been sensitive to terms of no more than second order in GM/rc^2, this quantity being no greater than 10^{-6}. In this chapter we are concerned with strong gravitational fields, $\frac{GM}{rc^2} \sim 1$.

We are already aware of astronomical objects in the vicinity of which such strong fields exist, although we have not as yet been able to study physical processes so close in. These objects are pulsars, so called because they emit pulses of radio waves (and in some cases visible light) with extreme regularity, the pulse periods lying in the range $\sim 50\,\text{ms}$ to 1 sec, [1]. The extreme regularity argues for a clock of stellar mass, while the short pulse period requires the dimensions of the emitting region to be less than about $c\tau$, 1500 km for $\tau = 50\,\text{ms}$. These objects are believed to be neutron stars, rotating at the pulse frequency and producing the pulses by the interaction of a magnetic dipole field with the surrounding plasma, [2]. Although the mechanism of formation of such neutron stars is not yet elucidated, it seems clear that they result from the gravitational collapse of a burnt out stellar core. When the core of a star has exhausted its nuclear fuel, thermal pressure cannot support it and if its mass exceeds $\sim 1.5\,M_\odot$ the pressure exerted by the Fermi sea of electrons cannot support it against gravity either, so that collapse is inevitable, squeezing the electrons into protons to form an immense nucleus consisting almost entirely of neutrons. The pulsar NP 0532, in the heart of the Crab nebula, itself the remnant of the supernova explosion of 1054 AD, is slowing down and the rate of decrease of rotational energy, assuming the pulsar to be indeed a neutron star, almost exactly matches the rate at which energy is radiated from the nebula.

At the surface of an object of mass $M_\odot$ and nuclear density, $3 \times 10^{14}\,\mathrm{gm\,cm^{-3}}$, the gravitational potential is

$$\frac{GM}{rc^2} \simeq 0.1$$

the surface radius being $\sim 10\,\mathrm{km}$.

In this chapter we shall only examine the properties of strong gravitational fields described by the Schwarzschild solution of the field equations. These fields are spherically symmetric and so correspond to a non-rotating source. The fields of rotating sources are of course of much greater astrophysical interest, but we are concerned here only with explaining what is meant by a hole and why it is black.

10.2 The propagation of light in strong fields

Consider the expression for the proper time interval in the Schwarzschild field, using standard coordinates:

$$d\tau^2 = \left(1 - \frac{2GM}{r}\right) dt^2 - r^2 \sin^2\theta\, d\varphi^2 - r^2\, d\theta^2 - \left(1 - \frac{2GM}{r}\right)^{-1} dr^2 \quad (9.5.3)(10.2.1)$$

The radial velocity of light is given in terms of the coordinates by setting $d\tau^2 = 0$ (and $d\varphi = d\theta = 0$)

$$\left.\frac{dr}{dt}\right|_{d\tau^2 = 0} = 1 - \frac{2GM}{r} . \qquad (10.2.2)$$

The Schwarzschild field of course applies to the empty space outside a spherically symmetric source. If such a source is concentrated within a coordinate radius $< 2GM/c^2$, then there exists a radius, the Schwarzschild radius,

$$R_s = \frac{2GM}{c^2}$$

at which the radial coordinate velocity of light is zero. An observer at infinite distance from the source has a clock running at the coordinate time rate: he 'sees' light passing the r markers at ever increasing intervals of his time, stopping altogether as $r \to R_s$. We must of course be careful about the interpretation of this result because we have not examined the significance of the radial coordinate.

The proper length of an infinitesimal measuring rod, oriented along the radius vector, at rest at radius r, is

$$d\ell^2 = g_{rr}\, dr^2$$

and so

$$dr = d\ell\sqrt{1 - \frac{2GM}{r}} .$$

As $r \to R_S$ the r markers are getting infinitely closely spaced.

We may evaluate the time taken for light to get out from radius r by integrating (10.2.2)

$$t = \int_{r_1}^{r_2} \frac{dr}{1 - \frac{2GM}{r}} = (r_2 - r_1) + 2GM\,\ell n\left(\frac{r_2 - 2GM}{r_1 - 2GM}\right) \tag{10.2.3}$$

and as $r_1 \to R_S$ this time $\to \infty$ for any $r_2 > r_1$.

If therefore we can use the Schwarzschild solution as $r \to R_S$ the conclusion is inescapable: light cannot get out past $r = R_S$ — and so neither can anything else capable of carrying a signal.

This result might appear to be associated with the singularity in g_{rr}. We see however that the rate at which a standard clock at rest runs is

$$d\tau = \sqrt{1 - \frac{2GM}{r}}\, dt$$

and so from any distance greater than R_S clocks appear to run infinitely slowly at $r = R_S$ and light is infinitely redshifted.

Let us examine these effects using isotropic coordinates, in which

$$d\tau^2 = \left(\frac{1 - \frac{MG}{2r}}{1 + \frac{MG}{2r}}\right)^2 dt^2 - \left(1 + \frac{MG}{2r}\right)^4 \left(dr^2 + r^2\, d\theta^2 + r^2 \sin^2\theta\, d\varphi^2\right) .$$

(9.6.5) (10.2.4)

The spatial piece of the metric tensor has no singularity in isotropic coordinates. However, the coordinate velocity of light goes to zero as $r \to R_I$, where

$$R_I = \frac{GM}{2c^2}$$

and the clocks stop at the same radius. In isotropic coordinates

$$d\ell = \left(1 + \frac{GM}{2r}\right)^2 dr$$

which behaves entirely sensibly except at $r = 0$. The time taken for light travelling radially is

$$t = \int_{r_1}^{r_2} dr \frac{\left(1 + \frac{GM}{2r}\right)^3}{1 - \frac{GM}{2r}} = (r_2 - r_1) - 2GM\,\ell n \frac{r_2}{r_1} + \left(\frac{GM}{2}\right)^2 \left(\frac{1}{r_2} - \frac{1}{r_1}\right) + 4GM\,\ell n\left(\frac{2r_2 - GM}{2r_1 - GM}\right) \tag{10.2.5}$$

which becomes infinite as $r_1 \to R_I$.

From the point of view of an observer located far from the source of the gravitational field, light emitted from a radius r is progressively redshifted and takes progressively longer to get out as r tends to the critical radius. For r less than or equal to the critical radius, a light signal cannot get out and a source of gravitation lying wholly within the critical radius is enveloped in an <u>event horizon</u> making communication with the outside world impossible. We may calculate how far away such an event horizon is from an external observer.

$$d\ell = \frac{dr}{\sqrt{1 - \frac{2GM}{r}}}$$

in standard coordinates, where $d\ell$ is the length of a real measuring rod stretched between coordinate markers labelled r and $r + dr$. Then

$$\ell = \int_{r_1}^{r_2} \frac{dr}{\sqrt{1 - \frac{2GM}{r}}} = \left[\sqrt{r^2 - 2\,r\,GM} + 2\,GM\,\ell n\left(\sqrt{r} + \sqrt{r - 2GM}\right)\right]_{r_1}^{r_2} . \tag{10.2.6}$$

As $r_1 \to R_S$ and $r_2 \gg GM$

$$\ell \to r_2 + GM\left(\ell n\left(\frac{2r_2}{GM}\right) - 1\right)$$

which is finite for finite r_2. In isotropic coordinates we find as $r_1 \to R_I$

$$\ell \to r_2 + GM\,\ell n\left(\frac{2r_2}{GM}\right)$$

the difference merely reflecting the relation

$$r_I = r_S + GM \quad \text{for} \quad r \gg GM .$$

We are not of course suggesting that a plumb line be lowered into such fields to make the measurement.

The existence of such one way event horizons is not unique to general relativity. Consider a rocket with accelerometer reading g. In an inertial frame its equation of motion is

$$\frac{v}{\sqrt{1 - v^2}} = gt \qquad (9.5.4) \tag{10.2.7}$$

and hence

$$x = \int_0^t \frac{gt\,dt}{\sqrt{1 + g^2t^2}} = \frac{1}{g}\left[\sqrt{1 + g^2t^2} - 1\right] . \tag{10.2.8}$$

A light pulse emitted at time t' from $x = 0$ is at $x_L = t - t'$ at time t. If $t' > \frac{1}{g}$ light can never catch up the rocket and so signals from space control cannot be received by the astronauts — space control lies behind an event horizon. The astronauts can nonetheless signal back to base and life goes on as usual in both space control and in the space vehicle. The signals start coming in again, of course, if the engines are switched off.

In the case of an event horizon surrounding a gravitating mass, there is no way of switching off the field producing the horizon. However, if the Schwarzschild metric and the equations of motion are correct even for r less than or equal to the critical radius, life goes on as usual in a space vehicle crossing this critical radius: the local physics is identical to that in a laboratory far removed from the source of the field, provided that the tidal forces may be neglected.

10.3 Particle motion in the field of a black hole [3]

Let us now consider the equation of motion of a particle falling inwards radially. Such a particle has zero angular momentum and so we can write Eq. (9.10.15) in the form

$$\left(\frac{dr}{d\tau}\right)^2 = \frac{1}{K^2} - \left(1 - \frac{2GM}{r}\right) \qquad (10.3.1)$$

where τ is the proper time on the trajectory, recorded by a clock on the falling particle. The quantity $\frac{1}{K^2} - 1$ clearly represents the square of the velocity as $r \to \infty$ (not the coordinate velocity at infinity) so

$$\left(\frac{dr}{d\tau}\right)^2 = v_\infty^2 + \frac{2GM}{r}$$

(compare the Newtonian equation $\frac{1}{2}mv^2 = \frac{1}{2}mv_\infty^2 + \frac{GMm}{r}$) and

$$d\tau = \frac{dr}{\left(v_\infty^2 + \frac{2GM}{r}\right)^{\frac{1}{2}}} \,. \qquad (10.3.2)$$

The proper time elapsing between r_2 and r_1 is thus

$$\tau = \int_{r_1}^{r_2} \frac{dr}{\left(v_\infty^2 + \frac{2GM}{r}\right)^{\frac{1}{2}}} = \frac{1}{v_\infty}\left\{\sqrt{r_2^2 + \frac{2r_2GM}{v_\infty^2}} - \sqrt{r_1^2 + \frac{2r_1GM}{v_\infty^2}} - \frac{2GM}{v_\infty^2}\ln\left[\frac{\sqrt{r_2} + \sqrt{r_2 + \frac{2GM}{v_\infty^2}}}{\sqrt{r_1} + \sqrt{r_1 + \frac{2GM}{v_\infty^2}}}\right]\right\}. \qquad (10.3.3)$$

This is finite for all r_1, even for $r_1 = 0$. A freely falling particle thus passes through the critical radius after a finite time and indeed arrives at the singularity at the origin, if accessible, after a finite proper time. The velocity $dr/d\tau$ increases continually in accord with the Newtonian formula. We must however emphasise that $dr/d\tau$ is the rate at which radial coordinate markers are passed measured with an on board chronometer. If instead of taking shipboard proper time as the time parameter we take coordinate time, which is also the proper time of an observer at infinity, we must use Eq. (9.10.16) and write

$$\left(\frac{dr}{dt}\right)^2 = \left(1-\frac{2GM}{r}\right)^2 - K^2\left(1-\frac{2GM}{r}\right)^3$$

$$= \left(1-\frac{2GM}{r}\right)^2\left[1-K^2\left(1-\frac{2GM}{r}\right)\right] \tag{10.3.4}$$

where $1-K^2$ represents the coordinate velocity squared at infinity, which is also the measured value at infinity. The velocity $\frac{dr}{dt} \to 0$ as $r \to 2GM$, as we would have expected since even the coordinate velocity of light goes to zero at the critical radius. We may evaluate the time taken to reach the critical radius:

$$t = \int \frac{dr}{\left(1-\frac{2GM}{r}\right)\sqrt{1-K^2\left(1-\frac{2GM}{r}\right)}}$$

$$= \frac{1}{\sqrt{1-K^2}}\int \frac{dr}{\left(1-\frac{2GM}{r}\right)\left(1+\frac{K^2}{1-K^2}\,\frac{2GM}{r}\right)^{\frac{1}{2}}} \tag{10.3.5}$$

Consider the two limits :

(i) $1-K^2 \to 1$, the test particle having the velocity of light. Then

$$t = r_2 - r_1 + 2GM\,\ell n\left(\frac{r_2-2GM}{r_1-2GM}\right) \tag{10.3.6}$$

which tends to infinity as $r_1 \to 2GM$.

(ii) In the limit $1-K^2 \to 0$ we have

$$t = \frac{1}{\sqrt{2GM}}\left\{\frac{2}{3}r^{\frac{3}{2}} + 4GM\,r^{\frac{1}{2}} + (2GM)^{\frac{3}{2}}\,\ell n\left(\frac{\sqrt{r}-\sqrt{2GM}}{\sqrt{r}+\sqrt{2GM}}\right)\right\}_{r_1}^{r_2} \tag{10.3.7}$$

which again tends to infinity as $r_1 \to 2GM$.

Thus from the point of view of an outside observer a particle approaching the critical radius slows down and takes an infinite time to reach it. From the point of view of an observer falling with the particle the critical radius is passed after a finite time and life goes on as usual. From the point of view of an outside observer, then, anything which passes the critical radius, using the proper time of the falling object as the time parameter, is effectively lost for ever and we shall now show that this is the inexorable fate of a large class of falling objects. It is most convenient to use for this study the

Eq. (9.10.15): we stick to the standard form of the Schwarzschild solution because of the simplicity of the equations of motion in these coordinates and because of the close analogy with the familiar equations of Newtonian theory. Write

$$\left(\frac{dr}{d\tau}\right)^2 = \frac{1}{K^2} - \left(1 - \frac{2GM}{r}\right) - \frac{H^2}{r^2}\left(1 - \frac{2GM}{r}\right) \tag{10.3.8}$$

with

$$r^2 \frac{d\varphi}{d\tau} = H .$$

Differentiating with respect to τ we obtain

$$\frac{d^2r}{d\tau^2} = \frac{H^2}{r^3} - \frac{GM}{r^2} - \frac{3GM}{r^4} H^2 . \tag{10.3.9}$$

Equation (10.3.8) is the analogue of the Newtonian equation for conservation of energy and Eq. (10.3.9) the analogue of the Newtonian equation for the radial acceleration. We interpret Eq. (10.3.8) in the following way. As $r \to \infty$ we find

$$\left(\frac{dr}{d\tau}\right)^2 \to \frac{1}{K^2} - 1$$

where

$$\left(\frac{dt}{d\tau}\right)^2 \to \frac{1}{K^2} .$$

Since

$$p_\mu = m_0 \frac{dx_\mu}{d\tau} ,$$

where m_0 is the proper mass, we identify the quantity $1/K^2$ with the square of the total energy at infinity for unit proper mass, E_∞. Then

$$\left(\frac{dr}{d\tau}\right)^2 = E_\infty^2 - \left(1 - \frac{2GM}{r}\right)\left(1 + \frac{H^2}{r^2}\right) . \tag{10.3.10}$$

In the Newtonian limit we replace τ by t, and replacing E_∞ by $(1 + T_\infty)$ find

$$\left(\frac{dr}{dt}\right)^2 = 2\,T_\infty - \frac{H^2}{r^2} + \frac{2GM}{r}$$

where T_∞ is the kinetic energy for unit mass and we have dropped those terms which are small in the Newtonian limit.

From Eq. (10.3.9) it is clear that we may regard the quantity

$$\frac{H^2}{2r^2}\left(1 - \frac{2GM}{r}\right) - \frac{GM}{r}$$

as the effective potential determining r as a function of τ. However, in relativistic problems we work with total energy more often than kinetic energy and it is convenient to write

$$\left.\begin{aligned} \left(\frac{dr}{d\tau}\right)^2 &= E_\infty^2 - V^2(r) \\ \text{where} \quad V^2(r) &= \left(1 - \frac{2GM}{r}\right)\left(1 + \frac{H^2}{r^2}\right) \end{aligned}\right\} \tag{10.3.11}$$

E_∞ is the total energy for unit mass at infinity (and need not be greater than unity) and H is the angular momentum for unit mass.

This potential function $V^2(r)$ has a long range attractive piece varying as r^{-1}, a shorter range repulsive piece varying as r^{-2} and an even shorter range attractive piece varying as r^{-3}. At very large r, $V^2(r) \sim 1$. As r decreases it will go below 1 and then for a sufficiently large value of H reach a minimum and increase with decreasing r. This increase is just the centrifugal repulsion, and these are precisely the characteristics encountered in the Newtonian problem. However, at very small r a centrifugal attraction prevents $V^2(r)$ from increasing to an infinite value at $r = 0$, produces a maximum in $V^2(r)$ and then digs a hole in the middle. We may note that a centrifugal attraction is present even in the equations of motion we obtained in the weak field case (Eqs. (5.4.14); (7.3.2)): we were forced to introduce such a term to account for the deflection of light by the Sun.

The maxima and minima of the potential function are given by

$$\frac{\partial V^2(r)}{\partial r} = 0 \quad \text{or} \quad \frac{H^2}{r^3}\left(1 - \frac{3GM}{r}\right) - \frac{GM}{r^2} = 0 \,. \tag{10.3.12}$$

The trivial solution of this equation is of course $r = \infty$. The other two solutions are

$$\frac{2}{r} = \frac{1}{3GM} \pm \sqrt{\frac{1}{(3GM)^2} - \frac{4}{3H^2}} \,. \tag{10.3.13}$$

There is neither a maximum nor a minimum if

$$\frac{4}{3H^2} > \frac{1}{9\,G^2M^2} \,, \qquad H < 2\sqrt{3}\,GM$$

At this value the maximum and minimum coalesce into a point of inflexion at $r = 6GM$ which represents the last possible stable circular orbit. Thus for any value of the energy at infinity there is no repulsion to stop a test particle with $H < 2\sqrt{3}\,GM$ from falling into $r = 2GM$ and hence being captured by the source of the gravitational field. In the Newtonian case of course any non-zero angular momentum is sufficient to prevent a test particle falling in to $r = 0$.

For a specified value of $H > 2\sqrt{3}\,GM$ the maximum of the centrifugal barrier occurs at r_{max}:

$$\frac{1}{r_{max}} = \frac{1}{6GM}\left\{1 + \sqrt{1 - 12\left(\frac{GM}{H}\right)^2}\right\} \,. \tag{10.3.14}$$

If the total energy is sufficient for the test particle to cross the barrier, then it will be swallowed up. The condition for capture is

$$E_\infty^2 > V^2(r_{max}) \ . \tag{10.3.15}$$

For a particle coming in from infinity we can express the angular momentum for unit mass in terms of an impact parameter and the energy at infinity. Far from the source, motion is approximately straight line, and so

$$r \cos \varphi = b \quad , \quad r \sin \varphi = x$$

where φ is the azimuthal angle measured from $\underline{b}$, whence

$$r^2 \frac{d\varphi}{d\tau} = b \frac{dx}{d\tau} = H$$

$$\left(\frac{dx}{d\tau}\right)^2 + 1 = E_\infty^2$$

so that

$$H^2 = b^2 (E_\infty^2 - 1) \ . \tag{10.3.16}$$

As $E_\infty \to \infty$ (corresponding to a particle moving with the velocity of light) the condition (10.3.15)

$$E_\infty^2 = V^2 (r_{max})$$

becomes

$$\frac{b^2}{r_{max}^2} \left(1 - \frac{2GM}{r_{max}}\right) = 1 \ .$$

With $r_{max} = 3GM$ we have

$$\pi b^2 = 27 \pi (GM)^2 \tag{10.3.17}$$

and this is the capture cross-section of an object of mass M, sufficiently compact to be enveloped by an event horizon, for relativistic particles. It is also clear that there is no stable orbit for relativistic particles, but that an unstable circular orbit exists at $r_{max} = 3GM$.

For non-relativistic particles starting at infinity

$$H^2 = b^2 \left\{\left(1 + \frac{v_\infty^2}{2}\right)^2 - 1\right\} \simeq v_\infty^2 b^2$$

of course. The energy condition is obtained by setting $E_\infty^2 - 1 \approx 0$ when we have

$$\frac{H^2}{r_{max}^2} = \frac{2GM}{r_{max}} + \frac{H^2}{r_{max}^2} \frac{2GM}{r_{max}} \ . \tag{10.3.18}$$

Set

$$\frac{1}{r_{max}} = \frac{\alpha}{GM}$$

and solve for α, obtaining

$$\frac{1}{r_{max}} = \frac{1}{4GM} \left\{1 + \sqrt{1 - 16 \left(\frac{GM}{H}\right)^2}\right\}$$

Compare with the condition determining r_{max} from the maximum of the potential (10.3.14)

$$\frac{1}{r_{max}} = \frac{1}{6GM}\left\{1 + \sqrt{1 - 12\left(\frac{GM}{H}\right)^2}\right\}$$

and so find

$$H = 4GM = v_\infty b\ .$$

Then

$$\pi b^2 = 16\,\pi\frac{(GM)^2}{v_\infty^2}\,. \qquad (10.3.19)$$

An object sufficiently condensed so as to lie wholly within its critical (Schwarzschild) radius thus exhibits the following features. There exists no stable circular orbit inside $r = 6GM$, and any test particle endowed with insufficient angular momentum irretrievably falls into the critical radius, taking infinite time (as seen by an external observer) to get there. An external observer cannot meaningfully ask what happens after this radius is passed. It is in this sense that such a collapsed object constitutes a hole in space. Light cannot escape through the critical radius and as this radius is approached light becomes infinitely redshifted. The hole is black.

10.4 The search for black holes

It may be that black holes were formed in the early stages of the evolution of the universe, and it may be that very massive black holes have been formed as the result of the implosion of star clusters or the cores of galaxies. We shall however limit ourselves to a brief discussion of the way in which black holes formed by the collapse of stars might manifest themselves [4].

A young star consists mostly of hydrogen. In the early stages of star formation a cloud of interstellar gas and dust collapses, heats up and at the same time radiates away the surplus energy released in contraction. When the interior becomes hot enough for the nuclear reactions which transmute hydrogen to helium to proceed, despite the inhibiting effect of the Coulomb repulsion, the star enters a quasi-equilibrium phase. In this phase the energy radiated from the hot surface is balanced by the nuclear energy sources in the core. Once the hydrogen in the core is exhausted, collapse of the core starts which is stopped when helium is burned to carbon and oxygen. Most of the available nuclear energy has been consumed in the formation of helium and so a massive core will evolve quickly towards iron, after which no further nuclear energy is available. If this core is sufficiently dense that the electrons are largely relativistic, the pressure exerted by this Fermi sea of electrons is insufficient to stop further collapse. The critical mass for this is known as the Chandrasekhar limit and is about $1.5\,M_\odot$.

If the degenerate core never rises above this limit, the electrons remain non-relativistic, the Fermi pressure prevents collapse and the star cools through a white dwarf phase in which the kinetic energy of the nuclei leaks away. If the core is above the Chandrasekhar limit, collapse continues in an unstable situation, the zero point energy of the electrons is raised to the level at which they are removed through the reaction

$$p + e^- \rightarrow n + \nu$$

and collapse only stops when the neutron pressure balances the gravitational pressure. The density is now nuclear density. The minimum mass for a neutron star is $\approx 0.1\,M_\odot$. (It is amusing but rather fortuitous that this result may be obtained merely by augmenting the semi-empirical formula for nuclear masses by a gravitational energy term.)

If we take the density of such a neutron star to be 3×10^{14} gm cm^{-3}, the radius for $1\,M_\odot$ is 10 km and $\frac{GM}{r} \approx 0.1$. Keeping the density constant and increasing the mass, $\frac{2GM}{r} \approx 1$ is reached at a mass of $\approx 4\,M_\odot$.

These calculations only give a very rough idea of the mass for which a neutron star surrounds itself with an event horizon. It is clear that in order to make a proper calculation it is necessary to put in a realistic equation of state for the neutronic matter and calculate the equilibrium conditions using general relativity instead of Newtonian gravitation [5]. A neutron star of mass less than $\approx 0.1\,M_\odot$ will expand to become a white dwarf: a neutron star of mass greater than a few $M_\odot$ will itself be unstable against gravitational collapse. We expect therefore the collapse of a stellar core with mass exceeding a few $M_\odot$ to lead to a black hole.

In searching for a black hole representing the end point of the evolution of a star we should therefore look for an object of several solar masses or more which is invisible and manifests itself only through its long range gravitational field. If we can find direct evidence of small size (remembering that $GM_\odot \approx 1.5$ km) and strong gravitational field, so much the better. It seems hopeless to search for isolated black holes, but some 85% of all stars are members of multiple systems [6]. The search has therefore concentrated on binary systems with one apparently normal member. Observation of the characteristics of the orbit of the primary can yield the mass of the unseen secondary. An example is the peculiar eclipsing binary ϵ Aurigae. The primary has an orbital velocity of 14 km s^{-1} (obtained from Doppler shifts) and the period of the system is 27.1 years (from both Doppler shifts and the eclipses). The velocity of the primary and the period give the mass of the invisible secondary in terms of the primary mass. The latter may be estimated from the visual appearance: it seems that the visible star is rather far advanced in its evolution and has a mass in the range $\approx 12\text{–}25\, M_{\odot}$, implying a secondary mass in the range $\approx 12\text{–}18\, M_{\odot}$, which is certainly far too massive for a neutron star. However, the eclipse lasts for about 400 days and if the secondary is compact it must be surrounded by a disc of fairly opaque muck which actually does the eclipsing. While such a model explains the light curve of the eclipses, we cannot conclude that ϵ Aurigae contains a good black hole candidate: it is possible that the secondary is simply a bit too faint to be visible [7].

The best candidate to date is the X-ray source Cygnus X-1. The visible star (HDE 226868) has all the hallmarks of a supergiant with mass 20–30 $M_{\odot}$ and from the Doppler shifts the velocity of the primary is $\approx 75\,\mathrm{km\,s^{-1}}$, [8] and the period is 5.6 days. The mass of the secondary is inferred to be $\geqslant 5\, M_{\odot}$. In this system there is evidence from the properties of the X-ray source that the secondary is compact [9]. First, there are strong irregular fluctuations with a timescale of ~ 0.1 sec. and irregular flickering on a millisecond time scale [10]. The X-rays are therefore produced from one or more regions only a few hundred kilometres across. Secondly, most of the X-ray energy radiated lies in the range 10–100 KeV. The X-rays are thought to result from the distended envelope of the supergiant streaming across the equilibrium point between the two gravitational fields and falling in towards the secondary to form a turbulent accretion disc, heating up and radiating [9]. If $m\,\mathrm{gm\,s^{-1}}$ fall in to radius r before getting hot, then very roughly

$$\mathcal{L}_x \approx \frac{GMm}{r} \approx \sigma r^2 T^4 \tag{10.4.1}$$

where M is the mass of the secondary, $\mathcal{L}_x$ the X-ray luminosity and T the temperature of the emitting region. The quantity σ is Stefan's constant. The X-ray luminosity is obtained from the X-ray intensity at the receiver and estimates of the distance of Cygnus X-1. The distance may be obtained from the brightness of the primary and the characteristics of the spectrum. This is a bit dangerous because the primary in such a system might be abnormal: a further method of estimating the distance is from the reddening of the light from the primary by interstellar dust [11]. The best estimate of distance is ~ 6000 light years and the corresponding value of $\mathcal{L}_x$ is $10^{37}\,\mathrm{erg\,s^{-1}}$ (about $10^3\, \mathcal{L}_{\odot}$).

If we take $T \sim 10^7\,{}^{\circ}\mathrm{K}$ (corresponding to a thermal energy in the gas orbiting the secondary of ~ 1 keV) then

$$r \approx 100 \text{ km}$$

and if $M = 6 M_\odot$ then

$$m \approx 10^{17} \text{ gm s}^{-1}$$
$$\approx 10^{-9} M_\odot$$

each year. This is a very reasonable rate of transfer: it must be remembered that the cooler outer regions of the disc of gas must not be too dense or the X-rays could not get out.

The X-ray source Hercules X-1 provides a beautiful illustration of this mechanism at work [9]. This binary system has a period of 1.7 days and the X-rays are pulsed with a period of just over 1.2 sec; in the range of (radio) pulsars. This pulsation presumably reflects the rotation of a neutron star, and has allowed the extraction of the velocity of the secondary through the periodic Doppler shift exhibited by the pulsing: it is 169 km s^{-1}, [12]. Optical Doppler studies of the primary (HZ Herculis) yield a primary velocity of ≈ 80 km s^{-1}, [13].

From these numbers we can work out at once the masses of both components. Assuming small eccentricity, as indicated by the details of the Doppler shifts, we have

$$\frac{M_1 v_1^2}{r_1} = \frac{G M_1 M_2}{(r_1 + r_2)^2} , \quad \frac{M_2 v_2^2}{r_2} = \frac{G M_1 M_2}{(r_1 + r_2)^2} \tag{10.4.2}$$

where r_1, r_2 are the distances from the centre of mass of the system. We also have

$$\frac{r_1}{v_1} = \frac{r_2}{v_2} = \frac{\tau}{2\pi} \tag{10.4.3}$$

where τ is the period. Then

$$\begin{aligned} M_1 &= 3.6 \times 10^{33} \text{ gm} \quad (1.8\ M_\odot) \\ M_2 &= 1.8 \times 10^{33} \text{ gm} \quad (0.9\ M_\odot) . \end{aligned} \tag{10.4.4}$$

The mass of the primary thus extracted is in agreement with the appearance of the primary HZ Herculis and the mass of the secondary is indeed right in the middle of the range of masses expected for neutron stars.

The distance of Hercules X-1 is $\approx 10^4$ light years and the X-ray luminosity is $\approx 10^{37}$ erg s^{-1}. This eclipsing system has a further amusing feature. The primary is brighter and bluer on the side of the X-ray source, becoming dimmer and redder when the primary is eclipsing the secondary, [14]. The atmosphere is presumably heated by the impact of X-rays from Hercules X-1 .

The X-ray source Centaurus X-3 has similar features [9]. The period is 2.1 days and the X-rays are pulsed with a period of 4.84 seconds: presumably the accreting secondary is another neutron star.

We cannot of course conclude that Cygnus X-1 contains a black hole. It is not impossible that the accreting object is less massive than $5\,M_\odot$. For example, perhaps it is a neutron star in orbit about a star of mass $5\,M_\odot$, which in turn is orbiting the primary, although this seems unlikely. However, even if the accreting mass is $\sim 5\,M_\odot$, all the activity the X-ray telescopes are looking at is occurring a hundred kilometres out and the hypothetical critical radius is at $\sim 15\,\mathrm{km}$. The critical radius which is the characteristic distinguishing the fields of black holes from other sources of strong gravitational fields is thus, at least for the moment, beyond observation.

10.5 The universe

On a large scale the universe is of roughly constant density. If it is both big enough and dense enough the gravitational field will make escape of light to infinity impossible and the universe would then be closed. For a proper discussion of cosmology it is necessary to use the solutions of Einstein's field equations in the presence of a uniform density of matter [15]: here we merely note that the condition the universe be closed is very approximately

$$\frac{GM_u}{R_u c^2} \approx 1. \tag{10.5.1}$$

The radius of the universe should be taken as $\approx 10^{28}$ cm, corresponding to expansion at approximately the velocity of light for about 10^{10} years. In order to satisfy Eq. (10.5.1) we need a density $\rho \approx 10^{-29}\,\mathrm{gm\,cm^{-3}}$. However the best estimates of the density currently available yield $\rho \approx 10^{-31}\,\mathrm{gm\,cm^{-3}}$, and it seems that the left-hand side of Eq. (10.5.1) is at least an order of magnitude less than unity, implying an open universe rather than a universal black hole, [16].

References

[1] For a review, see A. Hewish, Ann. Rev. Astronomy and Astrophysics, 8, 179 (1970).

[2] For a review, see A.G.W. Cameron, Ann. Rev. Astronomy and Astrophysics, 8, 265 (1970).

[3] General reference: C.W. Misner, K.S. Thorne and J.A. Wheeler, 'Gravitation', (Freeman 1973), Chapter 25.

[4] See D.D. Clayton, 'Principles of Stellar Evolution and Nucleosynthesis', (McGraw-Hill 1968); for an elementary discussion of stellar evolution, see M.G. Bowler, 'Nuclear Physics', (Pergamon 1973) Chapter 5.

[5] See S. Weinberg, 'Gravitation and Cosmology', (Wiley 1972) Chapter 11.
Ya.B. Zeldovitch and I.D. Novikov, 'Relativistic Astrophysics', vol.1, (University of Chicago 1971) Chapter 10.

[6] C.W. Allen, 'Astrophysical Quantities', (London 1973).

[7] A.G.W. Cameron, Nature, 229, 178 (1971).
R. Stothers, Nature, 229, 180 (1971).
P. Demarque and S.C. Morris, Nature, 230, 516 (1971).
A.G.W. Cameron, Nature Physical Science, 231, 148 (1971).

[8] R. Brucato and J. Kristian, Ap. J., 179, L129 (1973).

[9] For a review of compact X-ray sources, see
G.R. Blumenthal and W.H. Tucker, Ann. Rev. Astronomy and Astrophysics, 12, 23 (1974).
Two relevant semipopular articles are
'The Search for Black Holes', K.S. Thorne; Scientific American, 231, (6), 32 (1974).
'X-ray Emitting Double Stars', H. Gursky and E.P.J. van den Heuvel; Scientific American, 232, (3), 24 (1974).

[10] R.E. Rothschild et al., Ap. J., 189, L13 (1974).

[11] J. Bregman et al., Ap. J., 185, L117 (1973).

[12] H. Tananbaum et al., Ap. J., 174, L143 (1972).

[13] D. Crampton and J.B. Hutchings, Ap. J., 178, L65 (1972).

[14] J.N. Bahcall and N.A. Bahcall, Ap. J., 178, L1 (1972).

[15] See for example:
H.A. Atwater, 'Introduction to General Relativity', (Pergamon 1974) Ch.7.
S. Weinberg, 'Gravitation and Cosmology', (Wiley 1972) Part V.
C.W. Misner, K.S. Thorne and J.A. Wheeler, 'Gravitation', (Freeman 1973) Part VI.

[16] J.R. Gott et al., Ap. J., 194, 543 (1974).

Some semi-popular articles not mentioned in the text are:

I. Iben, 'Globular Cluster Stars', Sci. Am., 223, (1), 26 (1970)

J.P. Ostriker, 'The Nature of Pulsars', Sci. Am., 224, (1), 48 (1971)

M.A. Ruderman, 'Solid Stars', Sci. Am., 224, (2), 24 (1971).

C.M. Will, 'Einstein on the Firing Line', Physics Today, 25, (10), 23 (1972).

C.M. Will, 'Gravitation Theory', Sci. Am., 231, (5), 24 (1974).

K.S. Thorne, 'Gravitational Collapse', Sci. Am., 217, (5), 88 (1967).

R. Ruffini and J.A. Wheeler, 'Introducing the Black Hole', Physics Today, 24, (1), 30 (1971).

R. Penrose, 'Black Holes', Sci. Am., 226, (5), 38 (1972).

INDEX

OTHER TITLES IN THE SERIES IN NATURAL PHILOSOPHY

Vol. 1. DAVYDOV—Quantum Mechanics
Vol. 2. FOKKER—Time and Space, Weight and Inertia
Vol. 3. KAPLAN—Interstellar Gas Dynamics
Vol. 4. ABRIKOSOV, GOR'KOV and DZYALOSHINSKII—Quantum Field Theoretical Methods in Statistical Physics
Vol. 5. OKUN'—Weak Interaction of Elementary Particles
Vol. 6. SHKLOVSKII—Physics of the Solar Corona
Vol. 7. AKHIEZER *et al.*—Collective Oscillations in a Plasma
Vol. 8. KIRZHNITS—Field Theoretical Methods in Many-body Systems
Vol. 9. KLIMONTOVICH—The Statistical Theory of Non-equilibrium Processes in a Plasma
Vol. 10. KURTH—Introduction to Stellar Statistics
Vol. 11. CHALMERS—Atmospheric Electricity (2nd Edition)
Vol. 12. RENNER—Current Algebras and their Applications
Vol. 13. FAIN and KHANIN—Quantum Electronics, Volume 1—Basic Theory
Vol. 14. FAIN and KHANIN—Quantum Electronics, Volume 2—Maser Amplifiers and Oscillators
Vol. 15. MARCH—Liquid Metals
Vol. 16. HORI—Spectral Properties of Disordered Chains and Lattices
Vol. 17. SAINT JAMES, THOMAS and SARMA—Type II Superconductivity
Vol. 18. MARGENAU and KESTNER—Theory of Intermolecular Forces (2nd Edition)
Vol. 19. JANCEL—Foundations of Classical and Quantum Statistical Mechanics
Vol. 20. TAKAHASHI—An Introduction to Field Quantization
Vol. 21. YVON—Correlations and Entropy in Classical Statistical Mechanics
Vol. 22. PENROSE—Foundations of Statistical Mechanics
Vol. 23. VISCONTI—Quantum Field Theory, Volume 1
Vol. 24. FURTH—Fundamental Principles of Theoretical Physics
Vol. 25. ZHELEZNYAKOV—Radioemission of the Sun and Planets
Vol. 26. GRINDLAY—An Introduction to the Phenomenological Theory of Ferroelectricity
Vol. 27. UNGER—Introduction to Quantum Electronics
Vol. 28. KOGA—Introduction to Kinetic Theory: Stochastic Processes in Gaseous Systems
Vol. 29. GALASIEWICZ—Superconductivity and Quantum Fluids
Vol. 30. CONSTANTINESCU and MAGYARI—Problems in Quantum Mechanics
Vol. 31. KOTKIN and SERBO—Collection of Problems in Classical Mechanics
Vol. 32. PANCHEV—Random Functions and Turbulence
Vol. 33. TALPE—Theory of Experiments in a Paramagnetic Resonance
Vol. 34. TER HAAR—Elements of Hamiltonian Mechanics (2nd Edition)
Vol. 35. CLARKE and GRAINGER—Polarized Light and Optical Measurement
Vol. 36. HAUG—Theoretical Solid State Physics, Volume 1
Vol. 37. JORDAN and BEER—The Expanding Earth
Vol. 38. TODOROV—Analytical Properties of Feynman Diagrams in Quantum Field Theory